Future Tech

Future Tech

How to Capture Value from
Disruptive Industry Trends

Trond Arne Undheim

KoganPage

Publisher's note

Every possible effort has been made to ensure that the information contained in this book is accurate at the time of going to press, and the publishers and authors cannot accept responsibility for any errors or omissions, however caused. No responsibility for loss or damage occasioned to any person acting, or refraining from action, as a result of the material in this publication can be accepted by the editor, the publisher or the author.

First published in Great Britain and the United States in 2021 by Kogan Page Limited

2nd Floor, 45 Gee Street	122 W 27th St, 10th Floor	4737/23 Ansari Road
London	New York, NY 10001	Daryaganj
EC1V 3RS	USA	New Delhi 110002
United Kingdom		India

www.koganpage.com

Kogan Page books are printed on paper from sustainable forests.

ISBNs

Hardback   978 1 39860 034 8
Paperback 978 1 39860 032 4
Ebook   978 1 39860 033 1

British Library Cataloguing-in-Publication Data

A CIP record for this book is available from the British Library.

Library of Congress Control Number

2021930463

Typeset by Integra
Print production managed by Jellyfish
Printed and bound by CPI Group (UK) Ltd, Croydon CR0 4YY

CONTENTS

Introduction: What drives change

In the introduction, I chart the concept of change through a historical lens, tracing it back to the major thinkers on the industrial revolution, Marx, Weber, and Durkheim. I then turn to technological waves in a historical context, tracing them all the way to our contemporary technologies and concerns and through to coronavirus and what might come next. That becomes the backdrop for a quick tour of the structure of the book, which has two parts. The first part is dedicated to a macro perspective on technology. The second part is designed to help you, as an individual, respond to technological change, both as an individual and in the groups you find yourself in (at work, in social movements, in your social life, or in your family).

Change is a misunderstood concept. When weather, seasons, emotions, or people change, all of us are driven to oversimplify. We look for a single cause, even though we know that change is typically caused by a myriad of factors. Why? Because simplifying things helps us cope. Our psychological reaction precedes an intellectual explanation.

Technological change is particularly complicated. Historically, we tend to overvalue technology's role in change. That phenomenon even has a name—technological determinism. Even though this is a book called *Future Tech*, I will try not to fall victim to that determinism. Instead, I will go behind the technologies and look at what created them and what sustains them. Subsequently, I will get in front of them and chart what lies ahead, based on other equally salient disruptive forces, such as influences from government, business, society, or even the physical environment that surrounds us, Earth's ecosystem.

The future of technology is, of course, not an exact science. I share the fate of many futurists who have, wisely, stepped back a bit from prophecies. Instead, we chart scenarios. We model likely developments based on the forces of disruption we see in play already today. That can be done only by having a clear idea of how contemporary society is put together, a necessarily incomplete and simplified model of how things generally tend to work, which in turn requires an awareness of history.

To start with, let me just note that social change is usually equally important in terms of shaping technology as technology is in shaping social change. To prove that point, it is tempting to quickly begin to summarize the lessons of the 2020 coronavirus pandemic. That event indeed has set the context for an enormous amount of change—and it came from the environment (an animal carried the virus), to society (manifesting itself first in China, then in Italy, then across Europe, and in Iran, spreading to the United States, and then to Brazil), and only subsequently influencing sci-tech (stimulating a massive vaccine effort among all the world's top life science labs combined).

However, an even better perspective is gained from looking a bit further back in history. I would like to bring your attention to the industrial revolution. The reason I do that is that even though we call it an "industrial" revolution, the emphasis is often on technology when, in fact, it was social upheaval in cities that created the incentive and opportunity for such massive changes to take hold.

What the industrial revolution taught us about today's changes

The industrial revolution of the 18th century caused migrations, upheaval, and economic progress, and created new winners and losers, individually and collectively. German thinkers Karl Marx and Max Weber, and French sociologist Emile Durkheim, and others, each attempted to explain this change—Marx and Weber found the driving forces to be at the individual level, Durkheim found them mostly to be at the collective level. Both ways of looking at it can be fruitful.

Marx's observations on change

Marx's (1990) explanation, largely based on UK data, focused on how individuals respond to the class struggle. He held that class struggle (within the capitalist system) was inherent to the new industrial production of goods. A worker's identity was tied to the goods that were produced. To realize their true selves, workers had to rise above their status as servants to the technology to become owners of the instruments of production.

In other words, technological progress had to go hand in hand with changes in the ownership of the instruments of production. In Marx's mind, progress stems from the conflicts that necessarily arise out of people exercising their true interests. Marx's writings are commonly misunderstood as calls to action instead of explanations of the process of social and technical change, yet they were both. Any true revolutionary has a deep understanding of the society they are trying to change. Anything else would be unethical and short-sighted.

Weber's observations on change

Weber's (1922) explanation of change centered on the specific role of the Calvinist interpretation of the protestant work ethic in fueling capitalism as the answer to whether people should expect to be saved. Calvinists first broke from the Roman Catholic Church in the 16th century. Weber maintained that a complex set of rationales could help people attain meaning through the external evidence they saw from their efforts. As a result, hard work, discipline, and frugality became traits that, over time, produced tremendous change in the living conditions for workers as well as, in turn, for society. Weber's analysis, which appeared in 1904–05, was based on studying a particular Swiss-German version of Protestantism around the turn of the century, and in a historical perspective and context. Calvinism is arguably a major denomination still thriving today, although today's adherents, at least in the US, have abandoned many key Calvinist tenets and many Puritan ministers also have adapted the message slightly—although the work ethic certainly made it to the new world, and to America in particular. Today, however, it would be fair to say that hard work is motivated by the secular version of the American dream as much as by a religious motivation.

But Weber, in his foresight, had thought of that. He was also worried that rationalization would lead society astray. He worried about the inescapable iron cage of bureaucracy gone amok. Many have worried about bureaucracy and clearly there are governments that attempt to control individuals to an unnecessary degree.

What Weber pointed out was how the interplay of class, status, and party contributes to the distribution of power in a community. With that came the gradual understanding that economic value is complex and cannot just be derived from simply looking at the material conditions that created it. Weber, in fact, built the foundation for the understanding of social value, a much more complicated thing, and immensely much more valuable than pure economic gain. Technology, in this picture, is not an economic causality, it is much more complex, and is tied to social value and, in Weber's eyes, to the specific culture of the metropolis (Weber, 2005).

Durkheim's observations on change

Durkheim, on his side, focused on the role of the collective in bringing people together in new ways (enabling solidarity), such as in larger cities. To him, the collective was a more natural state (than simply being an individual) in which people were bonded by their affinities, to their families, friends, tribes, and co-workers. The collective represented a social consciousness that prioritized harmony over discord. Alienation, or suicide, was explained by the breakdown of the bonds of the collective, a function of the way society "worked," and not the result of individual action and thus subject to rationality or motivation, as such.

What gradually happened in the industrial revolution, according to Durkheim, was that the earlier *mechanic solidarity* (people mostly had shared beliefs and sentiments) gradually incorporated *organic solidarity* (further specialization and increased interdependence). As societies become more complex, the collective binds itself together in more intricate ways, but even more so feels the need to interact. There is a division of labor in which each part has its place.

The understanding of this process is not always perfect in any society. Note the surprise experienced by elites around the world during 2020's coronavirus pandemic when they discovered that food service workers, nurses, and delivery personnel are crucial to the operation of contemporary society. These are not merely frontline workers, they are "essential" workers, that is, they are key to the functioning of society as we know it.

Bourdieu on habits of change and stasis

Thinkers who followed in Durkheim's footsteps, notably fellow French sociologist Pierre Bourdieu (1977), have pointed out that the way change (and at times, lack thereof) occurs is through each collective (and social class, incidentally) learning to incorporate active habits (Bourdieu calls it *habitus*) from their surroundings through a socialization process. By the way, the bastardization of this line of thinking is found in today's applications of behavioral science to product development and marketing, mostly inspired by and loosely derived from empirical psychology. Purely utilitarian observations and experiments that attempt to prove which products various social groups want are, at best, shortcuts to a more complex reality, and at worst, provide pseudo-scientific rationales for consumption that scarcely has roots in human nature. Having said that, behavioral approaches may, on average, be slightly better than purely intuitive approaches that don't ground their observations in empirical behavior at all.

Conventional wisdom would indicate that children of academics tend to have deep discussions over the dinner table, read books on the bookshelf and dig into multiple sources online, and learn to learn, which is generically useful. Yet they may lack fundamental practical or commercial skills. Children of police learn to serve the community through prioritizing maintaining order, and typically follow a career path along those lines. Children of executives are taught about business by example, may learn to create material value, above all else, but they also, these days, tend to learn something about managing people, which is an essential skill. Children of engineers model another professional behavior, learn to take apart and put together machines, and so learn to create new, better machines, and they have a distinct advantage when it comes to understanding science and technology. Children of entrepreneurs learn to recognize challenges and embrace novelty and risk to solve them, and they learn the value of taking outsized risk (and they occasionally learn about entrepreneurial failure).

Now this would, again, border on determinism, if I did not insert that people can escape their upbringing (to some extent). Some children even make it a point to do things differently from their parents and may rebel for years, even for a lifetime. The point is simply that escaping your childhood is that much more difficult and kudos to those who feel they need it, want it, and achieve it. For instance, it has taken me near 50 years to realize that I am much more suited to life as an intellectual than life as a corporate executive, combined with being an entrepreneurial tinkerer and being a connector between worlds, even though that was already given from my upbringing. I think many of us resist our upbringing so much that we cannot see straight about all the upsides of following in our own footsteps, as it were, as opposed to seeing it as being forced into something by fate.

Learning also occurs across cohorts, among groups of children who spend enough time together to mirror each other's interests and so on. There is a whole strand of thinking along these lines called social learning. These habits, in turn, become structural tools (legacy behavior patterns) that define a social group's identity, and in so doing defend it from outsiders. These ways of behaving become "what really matters," regardless of the neuroplasticity that always remains inherent in you and could be mobilized if you tried harder.

Integrating strands of thought on change

Durkheim and Bourdieu are "functionalist" thinkers. To them, everything has a "goal," but the full extent of that goal is not simple, and not visible to each person. Contrary to what some people think, this line of thinking is very powerful in attempting to understand how people cope with sweeping change, because it illustrates how it's even possible to feel "okay" in the midst of it all.

Neither Weber nor Marx ever fully explained how an individual *copes* with their predicament. Marx, in fact, argued the answer always was to rebel, because he hyper-emphasized fairness. What Durkheim grasped, and has to teach us today, is that even though the function of something might be relatively static, its role can shift, even quite radically, without altering the basic understanding that "life goes on." This is a useful insight when trying to understand change, crisis or innovation. We can, in fact, simultaneously believe in collective values and take issue with some of the excesses of prioritizing the group over allowing for individual expression.

While Marx's, Weber's, and Durkheim's views differ as to whether technological progress had a greater impact on the individual or the collective, they would each agree that change impacts both the individual and the collective. It was difficult even for social scientist icons to keep the two apart. The process of change based on what occurred 200 years ago (the industrial revolution and its aftermath) is still poorly understood. What were the most important factors in kickstarting the changes? Why were some parts of the world industrialized much faster than other parts? Why did some individuals and social groups persist in trying to reject change instead of embracing it? If you accept that change, as such, and its social causes and consequences, are still shrouded in mystery, how can we expect to fully understand what's happening around us right now? The interpretive process has, for sure, not gotten easier with time because new variables are constantly added to the mix. History may not be a perfect guide, but rejecting its lessons does not do us any favors.

Technological waves in a historical context

The best answer to the mystery of change is that true understanding takes time and requires historical context. Which is why we now, several hundred years later, can start to appreciate how the historically specific *"waves" of technology* a few years later typically become viewed as *economic waves* because the effect is now visible in overall productivity. However, if we, as Marx, Weber, and Durkheim taught us, instead first (and ultimately) try to understand the effects in terms of *social waves*, we can better capture the gradual shift that also occurs in the way social dynamics can play out once technologies are part of society's infrastructure.

Not all technologies reach that level of maturity. For those that do, the importance of the platform created by the install base (e.g. the railway infrastructure) far outweighs the individual performance of the trains running on that infrastructure. This fact is quite frustrating for consumers who ask for faster trains, better cars, and more stable computers.

As consumers, we generally believe (in vain) that technologies are only about making things better, when in fact all they typically do is make things different, and usually—at least when successful—more standardized. The improvements that endure (when we are lucky, that is) are all due to putting technologies to use in an efficient manner, and for that we need to know where the train lines should go, what features will become important in cars, and what computers really are all about—are they productivity tools, entertainment devices, or control machines for government suppression? Or are they all of the above in certain situations?

These are much more complex questions than building the first prototype of each of these technologies, and something about which inventors, engineers, or vendors typically have very little to say and very little influence over. Historically, this was even more true, because the insight that has slowly emerged about user testing and design thinking was not prevalent in earlier technological waves. Thus, to some extent, we would have to consider the success of all those technologies as a stroke of luck.

Unbundling the second industrial revolution of the 20th century

Immediately after the second industrial revolution of the 20th century which brought us trains, electricity, and telephones, many observers initially thought that a *post-industrial dynamic* was at play, and that it was the service sector, not technology, that was going to drive all the changes. In 1976, Daniel Bell published *The Coming of Post-Industrial Society: A Venture in Social Forecasting*, stating that the rise of the service sector and a new class of professionals in the United States proved that technology had run its course.

In some ways, observers such as Bell argued "everything has been invented" and all we have to do is watch history play out, letting technology "service" society and servicing each other with technology as a tool to make it more efficient.

Technology's embedding with the constant struggles of democracy

On the political science side, this kind of argument is what led to Francis Fukuyama's book-length essay *The End of History and the Last Man* (1992), in which he claimed that western liberal democracy had won, and that there were no other models of government that would succeed. But as we know from the past decade, history didn't end, progress didn't stop, and democracy hasn't won. At least, there are many forms of democracy and some are only disguised as democracies for the time being. There are also substantial voices questioning the legitimacy of what many had considered unquestionably democratic institutions, processes, and electorate processes.

The current unraveling of the United States as a superpower and as a democratic governance entity, regardless of the causes, which we have no time to get into in this book, is a case in point. The rise of far-right movements in European politics is another. Things are never that simple. Manufacturing, in that view, in which one lumped in technology as simply a means of more efficient production, was simply the means to produce an end. We are now entering an era where we are slightly wiser, and where manufacturing will regain some of its stature as a legitimate industry in which essential functions are being carried out, but the process of gaining that awareness has been slow.

Knowledge, and human capital, became what mattered. Why knowledge? Because knowing what (insight) is going on is a big prerequisite for knowing how (technology). In this human-centric worldview, machines are mere tools, instruments for the human domination of all things. There was something comforting in that view. Again, we could focus on literature, on academia, on thinking as opposed to action, and on building skills needed to manage the technologies that were already developed.

Is knowledge the true source of invention and innovation, or are material changes and technology? Obviously, it depends on how wide you make that term. For a few decades, it almost looked like this was an accurate description. Few truly novel inventions arrived on the scene between 1930 and 1970, as I put the jet engine (1930) as happening right at the end of the past cycle of innovation and discount soft-serve ice cream (1938), ATMs (1939), superglue (1942), the microwave oven (1945), and the birth control pill (1960), and even—under doubt—freeze-dried coffee (1964). I will say that the fiber optic cable (1952) was pretty amazing, though. Having mentioned all those fly-in-the-face inventions, there were also under-the-radar innovation developments underway, but the point is this, they did not come to fruition in that era.

What was happening was not just society riding a technological wave and adjusting to them. In fact, many of the aforementioned innovations are somewhat insignificant in the greater picture—they are small spikes on the slow, ongoing innovation curve. Rather, it was a period dominated by a depression, followed by a world war, followed by immense rebuilding of society and corresponding economic progress.

But the technological change that fostered it was incremental, which does not at all mean that it was insignificant. It just was not disruptive. And because it was not disruptive, it could be exploited by the new class of professionals, technocrats who did not know the details of the technologies but who could serve as managers.

From this gist rose the whole cadre of professional managers in industry, as corporations grew and needed principles to manage their workers. Management practice fostered schools of management, and the MBA degree was born. The first MBA was awarded at Harvard in 1908 but only became a globally recognized degree by the 1950s. It was all very convenient and responsive to the changes that were happening in the workplace. But as we all know, history tends to get more complicated.

As infrastructure increasingly developed outside the government's immediate control, or was slowly deregulated in order to achieve what one thought of as higher efficiency, the emerging platforms consolidated, started to interact more closely, and then, seemingly "out of the blue," another technology wave hit us, starting in the mid-1970s, with GPS (1973) and magnetic resonance imaging, also known simply as MRI (1977), and one that had been brewing in the U.S. and U.K. military and in industrial laboratories for three decades: digitalization. At first, it was only a game of number crunching, and it happened on isolated machines. You can only go so far with that and the applications were important, such as statistics, cryptography, and weather prediction. But not they were not earth shattering at this point, partly because the use cases were limited and the full scope of those technologies had not yet been reached.

The consumer tech upheavals of the 1990s and 2000s

Then, in 1990, came the third industrial revolution and the introduction of the internet and online markets, enabled by the internet's ecommerce opportunities (1991) and the consumer release of WiFi (1997). With e-commerce and now blockchain, transactions can occur without central points of authority. Artificial intelligence has been applied to image recognition, chess game aptitude, and analytics. While we still have no idea what these technologies will lead to, we can and should speculate. Why? Because the advent of so many new technologies together holds enormous potential for cross-fertilization and wide-ranging, potentially uncontrollable, impacts.

Throughout the 2000s, there was a series of (arguably) great inventions and milestones, including the iPod (2001), which was the precursor to MP3 players, the first camera phone in various Asian phones by Sharp and Samsung and Nokia's N90 breakthrough 2005 phone with Zeiss optics, the final sequencing of human DNA from the human genome project (2003), voice calls over the internet with Skype (2003), social media with MySpace (2003) and Facebook (2004), YouTube (2005), Twitter (2006), the iPhone (2007), and the Amazon Kindle e-book reader (2007), to mention some (Forrest, 2015).

The environment strikes back: coronavirus of 2020

Occasionally, history brings an event that appears to come out of left field (it seldom does, it just looks that way at the time). This book is not the story of the impact of coronavirus. However, the virus will drastically affect the next decade's technology evolution. Arguably, the impact of the coronavirus pandemic will unleash a fourth industrial revolution, only a brief generation after the third, which might not have happened otherwise.

What that revolution will entail is still shaking out, but it clearly has to do with a deep societal response to the systemic risks presented by infectious disease run amok, and a set of social practices to counter the negative consequences of this increased risk. It also, hopefully, serves as the dress rehearsal for even worse pandemics and, to some extent, for the biggest crisis of them all, extinction-scale climate change. Furthermore, it seems clear that technology will play a part in enabling more mature and widespread usage of the technologies brought to us by the third industrial revolution. Finally, we are about to see the most severe split of haves and have nots in several hundred years: those who fully embrace, and are able to sustain, systemic risk and those who do not.

While we can look at previous industrial revolutions and reduce the changes that occurred to a single driving factor—say technology, politics, innovation, or consumer demands, take your pick—we need to overcome our inclination to oversimplify the explanation for these changes. While technology, politics, innovation, and even consumer demand are all important factors, each one of them is only a *contributing* factor. What drives change is not only skills, not only technologies, not only politics, or even human capital in and of itself. Rather, change is driven by the interaction of these four forces, and in the case of coronavirus, by the contextual force which is our embedding, our footprint, in the physical ecosystem around us, in this case, most likely, our proximity to wild animals (e.g. zoonotic spillover). To gain any insights, we need to watch each factor very carefully.

To understand technology, start analyzing what it was intended for

This book addresses the trends that both lead up to and follow from the emergence of new technologies. Rather than addressing technologies as such, we will tackle the evergreen principles that bring them about and modulate their usage, and what to do about it, for most of the book. Technology means "knowing *how*." Technology is just one factor and it is not always even the most crucial facet. Knowing how to do something doesn't help you figure out the motivation for using it (social), what makes it an improvement (innovation), or what safeguards are necessary to prevent against misuse (regulation).

The misguided notion of STEM

It is ironic how only a certain aspect of tech literacy, called STEM (the acronym used to push a certain basic knowhow in Science, Technology, Engineering and Math), has indeed become a fashionable basic skillset taught from K-12 to universities. As we will see, the focus on such a limited set of instrumental skills is alluring but shortsighted.

Very briefly, STEM is not what STEM is about. Truly appreciating STEM's potential requires in-depth awareness of the rationale, the meaning of technology, and its potential (desired and undesired) use cases. There is even the argument, as top engineering schools revise their curricula to meet the demands of today's students and what we now know in terms of learning science, that you cannot productively teach STEM aside from how it is used (NEET, 2020).

As with new technologies, the same goes for the "Big Data" that new technologies create. Data itself—even tons of it—does not yield better insights. We now make inexpensive sensors that can pick up data on just about anything from temperature to human emotion, but what are we looking to find or solve in all of this data? In science, reaching a conclusion too quickly and based on weak or faulty premises is called a fallacy (IEP, 2020). What we are dealing with here is typically the "fallacy of defective induction." Simply put, the availability of many observations does not necessarily make it easier to arrive at a theory that explains what is going on. An abundance of data simply creates an endless availability of potentially spurious correlations, things that co-exist in some sort of pattern that we far too quickly assign as a cause and effect. What we need is some understanding of what ties our world together.

The 21st century's challenges: what future tech is all about

In *Future Tech*, I ask a simple question: How should we, as business professionals, policy makers, and entrepreneurs, respond to the deeper tech trends that will shape our decade? I also ask you to do the research yourself and become enough of an insider to know what is likely to only be a fad, and what likely is a lasting, disruptive change you need to focus on. With technology, there is no substitute for depth. Do your homework. Develop your own informed opinion based on looking into the issues. Don't rely on random newsletters coming your way or annual emerging technology clickbait reports from consulting firms.

This book is about technologies that have direct relevance across many industries and across society in the 3–5-year range and typically through the decade. Many of them will foster lasting changes and become valuable platforms, and there is no knowing what innovations, business models, and social practices they will foster.

The main forces of disruption

The first part of the book is all about the macro-view on technology. How does it function in society? What do we know about how it works in various disruptive ways? Which technologies matter in this decade?

In Chapter 1, I explain the forces of disruption: technology, policy, business, social forces, and the environmental context that surrounds them. Tracking changes in emerging technologies, policy, business, and social dynamics can help you determine which priorities to chase, which interrelationships to be aware of, and what gaps to fill. While simple, this is a far better approach than simply chasing the hottest technologies and tracking whatever information you can find on the internet. In fact, the precision needed from digging at least one level deeper into each of these four categories of disruptive forces is where the true power of the analysis is revealed. That is why I have written a book about it and not just a chapter.

How sci-tech fosters innovation

In Chapter 2, I investigate how science and technology foster innovation, each in different ways. Three main lenses are helpful: platform technologies, taxonomies, and tech visualization.

Government policy regulates the playing field for tech

In Chapter 3, I look at the effect of government interaction in emerging markets that were created or are impacted by innovation. While many believe that government standardization, prohibitions, or detailed rules constrain and slow innovation, the truth is that by enabling a level playing field, government regulation benefits equal market access, consumer protection, and public safety, such as in the case of clinical trials for new drugs. Those who attempt to innovate within "regulated" industries without regard to regulations are likely to suffer setbacks even if their tech is good. The startup 23andMe is a case in point, as it miscalculated what a startup would be allowed to communicate to consumers about serious health issues—and the FDA noticed.

The rise of business models

We may not spend a lot of time thinking about them, but in Chapter 4 I reveal how business models have become a major instrument of innovation.

From being relatively unimportant because they were so commonplace and changed relatively slowly, the novel business models of today can reshape industries almost instantly. Perhaps the best meta-business model is that of creating a startup, and we will particularly look at a few unicorn startups, such as the sharing economy companies (Uber, Airbnb). However, the novel business model of today may not help us understand what will work tomorrow. Rather, to protect our business and to capture opportunity, we have to constantly look out for such novelty ourselves.

Social dynamics stems or stimulates tech

In Chapter 5, I emphasize the role of social dynamics in the adoption of trends. Social dynamics (i.e. whether people "catch on" or "adopt") can both preempt and stall all other forces by the sheer force of numbers. Consumers have become immensely powerful in retail and arguably have more influence over what is being sold than shops, vendors, or even B2C entrepreneurs. Whether this will last and what the future holds for ownership, pricing, product strategy, physical retail, and a host of other issues will be discussed in this chapter. I'll also consider the science behind psychographics, which was popularized by disgraced pollster/influencer Cambridge Analytica, but which is both a richer tradition than that saga contends and somewhat less powerful and scary than initially assumed.

The technologies of the decade

Chapter 6 dives into five technologies that matter at this time: artificial intelligence, blockchain, robotics, synthetic biology, and 3D printing. Why? They interact with each other, creating previously unthinkable conditions of technological, biological, material, social, and psychological change. Using these technologies, items as commonplace and well-known as a wall, a piece of cloth, and a human being may become nearly unrecognizable from their predecessors over the past 1,000 years within the span of this decade. How? Well, it will take us about 10 pages to explain. We quickly describe what each technology is capable of today, what it might achieve in 10 years and who you should track (scientists, innovators, startup founders), and what you should read (publications) or what tech conferences you should attend, to have any inkling about where it truly is evolving day by day, year by year.

Micro-view: how individuals respond to disruption

In the second part of the book, I focus on how individuals can best equip themselves to take advantage of these disruptions to their personal and professional lives in order to both understand and profit from change.

Polymaths of the 21st century

In Chapter 7, I outline the growth opportunities for individuals in this new landscape—specifically, how to become a "T-shaped expert", grow into far more useful "Pi-shaped" experts, or even move beyond expertise as such (which is only part of the ingredients of change) and into the territory of a polymath—fruitfully combining deep perspectives from several domains (which is the basis for recipes of change). The ideal is not only to be an expert in two or more domains and very deeply immersed in a dozen, if not more, but to demonstrate results from it that can benefit humanity. We cannot afford to have only an elite class of citizens perform this function. Rather, it has to become a widespread capability.

Personalizing your insight

In Chapter 8, I suggest a number of tools to help you tackle the future head on. There are many tools available to track trends, including low-hanging fruit (subscription to market research, consumer search engines), traditional approaches (relying on trade association newsletters and gatherings), time-consuming ones (attending industry events), expensive one-off stints (hiring strategy consultants), novelty tricks (on-demand expert networks), in-house approaches (strategy, internal consulting or R&D teams), or partnerships (universities, accelerators, VC firms, etc.). Given today's shifting knowledge needs, none of these works in isolation. The market for online content discovery tools will only continue to grow as the complexity of navigating online content increases. So as tools evolve, so should you. The winners of tomorrow will have access to a personalized growth toolkit. What does that mean? Simply put, you need a diversified approach to acquiring and processing information, deliberating decisions, and scaling your implementations. Whatever you do, do not go it 100 percent alone. You will surely fail.

A down-to-earth view on man/machine symbiosis

In Chapter 9, I suggest you take an ambitious stance and embrace the man/machine approach. Actively seek integration with various available machine intelligences and hardware. Make sure to spend enough time training the most advanced AI you can get hold of—everyone else who is smart will be doing the same. The emphasis of the next 3–5 years and likely the whole decade will be on augmenting human capability, not completely replacing it with AI-infused robotics.

The mixed blessings of change—human enhancement

I conclude by looking at the mixed blessings of change. All trends point to the fact that by 2030, unless we take action, the global labor force will be led by an elite set of knowledge workers enhanced and enabled by robotic AI—and by the next generation the roles will likely be reversed. What I mean is that you are likely looking at a scenario where there is an army of AIs (in robotic form factor or simply in software) and a smaller set of humans who (hopefully) control and enhance their work.

You can view this as a threat or as an opportunity, depending on how much you value going to work. However, beware that already in 1930, the American economist John Maynard Keynes predicted a society so prosperous that few had to work (Rosen, 2016). It turned out that increased productivity has not translated into increased leisure time, even for the elite (with some exceptions). Why, then, do we still work long hours? Certainly not awaiting the robots. We work because that is who we are. Our essential function as human beings is not likely to change anytime soon. We will just find other ways to work. Note that the Manufacturing Institute's skills gap study (2018) predicts some 2.4 million manufacturing jobs will be awaiting skilled workers between the time of the study and 2028.

The strategies for capturing surplus value from disruptive industry trends, which all tend to involve technology at this point, remain largely the same as they were before technology became a major factor. To capture value, you need to deeply understand what value is before it emerges and everyone else knows.

The question of (societal) value beyond pure economics

However, the entire question of value is much more complicated. Recently, as Wittenberg-Cox (2020) writes, a slew of female economists—Esther

Duflo, Stephanie Kelton, Mariana Mazzucato, Carlota Perez, and Kate Raworth—has begun to gain ground questioning what it actually means. Creating economic value is traditionally defined in monetary terms, but would there be a broader construct that could encompass how it serves people? Take the concept of gross domestic product (GDP), which ignores unpaid labor like housework and parenting and is sometimes antithetical to sustainable growth because it also finances weapons or fossil fuels. Finally, Wittenberg-Cox says, Carlota Perez reminds us how new growth is spurred by aspirational desires as opposed to guilt and fear.

Viewed that way, the future will not cause the angst that my fellow futurist, Alvin Toffler, captured in his book *Future Shock* (1970) 50 years ago. Even though the future has just begun, we have come a long way in our understanding of it—but that knowledge needs to be shared, again and again. If we keep learning, our technological future can be bright, both as individuals and as a collective, and it can serve the greater purpose of building a human future—a place where we can all thrive.

Key takeaways and reflections

1 Technological change is complex, yet we seek ways to simplify so we can understand it. How would you put the evolution of technology into your own words? Try to jot down a few technologies you care about and write a sentence or two about what you know about their origin.

2 Think about the first and second industrial revolutions and then think about what is happening today. Which period would you rather live in? In which period would you thrive? Which period will be viewed as more significant, looking back?

3 Consider the new framework, the forces of disruption. Can you describe each of the forces in your own words? Which of them are you most familiar with? Which one will it require more investment of time and effort to untangle?

4 Consider the notion of societal value compared with economic value and how each currently is measured or not measured. What human activities contribute true value? What is a fair measure of value?

The forces of disruption: Tech, policy, business models, social dynamics, and the environment

01

In this chapter, I explain the forces of disruption: technology, policy, business models, social forces, and the environmental context that surrounds them. Tracking changes in emerging technologies, trying to enact policy changes or predict regulatory changes, deeply understanding how organizations deploy various business models, and pondering social dynamics, such as the emergence and impact of social movements, as well as consumer activism, can help you determine which priorities to chase, which interrelationships to be aware of, and what gaps to fill. While it may sound too simple, and there is nothing magical about the forces of disruption framework (which is historically known by many names), it is quite complex to carry out as rapidly and in as many fields as needed, yet systematically and thoroughly. However, I promise you it is a far better approach than simply chasing the hottest technologies and tracking whatever information you can find on the internet. The chapter is an overview of the more detailed discussion to come in chapters tackling each of the forces separately. It also explains how they are interrelated by introducing the notion of a "living" entity I call the biosphere of innovation.

How to pick which technologies to focus on?

As an innovation matchmaker at MIT, I met with CEOs, strategists, innovation scouts, R&D directors, and business unit leaders from around the world. Most of them were interested in which specific technologies could impact their business. They usually came with a few they already had identified. That approach is sound, but I am sure most of them would agree that the way in which they came to their chosen few could always be improved. This book gives my view on how. My mission in this chapter is more to illustrate that no discussion of technology happens without picking a few technologies to focus on. Otherwise, it is hard to know what we are talking about since the concept is so wide.

For whatever reason, executives I meet typically name only 3–5 technologies they are looking at. Perhaps because it is hard to focus on more than that. Reducing complexity is a necessary and often effective communication strategy. It is also risky. Corporate (and human) priorities tend to come in odd numbers—three, five, seven at most. Once you get to 10 you are still making a shortlist, but deep tracking is likely unmanageable for most organizations unless they have an R&D lab to complement external scouting efforts. Once you boil down the focus to three technologies, you have something snappy, but typically at the price of missing aspects that already are, or will become, important to your business. A whole other matter is that it's not the technology that will make the difference but the awareness of the precise business challenge you are trying to address. There are often many ways to handle a specific challenge, and technology, while important, is only one ingredient.

How the myth of technology trends gets created

Technologies "change" each year, meaning the emphasis, the R&D progress, or even the potential use cases may expand or even contract. Despite this, the technologies that are mentioned by tech scouts across industries are remarkably similar in any given year. The industry discourse tends to closely follow what Gartner or Forrester or perhaps McKinsey or Deloitte, or occasionally the World Economic Forum or MIT Tech Review, or others, have put in their annual "trendy new technology" reports.

Each industry also tends to have its own favorite set of analyst firms, so there are a few more influences for each industry, but not much more. For example, in manufacturing, there is the Wohlers Report; in the internet VC investment industry, there is Mary Meeker's report; in blockchain there's Ryan Selkis'

Crypto Theses. These reports are immensely useful because they boil down what's happened over the past year into discernible trends and topics of interest.

However, what these companies also do in writing and promoting these reports is to create a myth that there is a set of technologies that matters (more) in any given year. But technology (and business) does not work like that. It is the marketing logic about getting attention around the work your research firm does that needs it and is willing to feed it to the market. My issue is not that it is not useful. It is. It is just not sufficient.

As much as I am a believer in simplicity and in reducing the noise, that worried me then and worries me now. Out of all that complexity, how can we reduce our focal lens to only three sources resulting in 5–7 things you focus on? In addition, the lens they would have had to apply includes which company they work for and which industry and specific sector they are in. The complexity is enormous, yet the result can be incapsulated in the term "AI." I do not think so.

Tech may be the answer but what is the question?

Part of the problem is the question asked, of course. When I change the question, the real challenge innovation executives and scouts have comes to the surface, which makes my job easier. My approach is always to get to the core challenge behind their question. To do so, I tend to ask them to narrow it down to one or two challenges. This is usually easy for them to do, and once the business challenge has been revealed (they are not always planning to reveal it), they remarkably quickly converge onto one technology that really is the focus of their CEO or leadership team, or ideally flesh out the actual underlying business challenge, whether it is related to efficiency or to developing new breakthrough products.

Questioning the tech buzzwords of 2016–2021

In 2016 it was all about the Internet of Things (IoT). In 2017 it was all about artificial intelligence. In 2018 it was about quantum computing. In 2019, we seemed infatuated by the promise of immersive experiences, autonomous driving, and the fear of too little cybersecurity. In 2020, for what it's worth, those immersive experiences are coming to life in the form of augmented reality based on breakthroughs in IoT and the wider availability of 5G mobile networks. For those who don't follow the buzzwords accurately, they might still talk about Big Data. For 2021, I will forecast supply chain issues (security, automation, robotics, 3D localization), online learning, and healthcare AI as candidate topics even before coronavirus entered the scene. After the pandemic, supply chain, augmented reality, online learning, and healthcare AI will move to the forefront.

Tech as understood by industry vs. tech as understood by academics

Tech discovery meetings between MIT staff and corporate executives illustrate a pitfall to be aware of: that of assuming the trends of the day can be aptly summarized by a trend report a consultant team creates in a few months of hectic work, looking at a myriad of open sources.

Contrast that with the fact that academics spend years or typically decades getting to the point where artificial intelligence or quantum computers advance from merely scientifically interesting to practically applicable. Although academics at top engineering schools have become more sophisticated at translating between academic and industrial concerns, they are not as aware of the industrial context of their own domains of knowledge. Communicating with an industry tech scout or a division head of technology, they need to be.

Sometimes, there is also a lack of respect between the two. The academic will assume industry does not know research but will overcompensate in trying to claim they understand industry's concerns. Industry might assume that the academic does not care too much about their commercial interests and might overcompensate by claiming their R&D labs are equally advanced on the tech side as the university lab (and in some rare cases they are right).

In what way is the judgment call on *when* a technology becomes useful somehow a simple decision? Also, industrial concerns sometimes influence academic work directly, so these consulting reports are often an unnecessary layer in between. Either way, there are at least three distinct ways of speaking about technology—the industry way, the consulting way, and the academic way. Add the journalistic perspective (even science journalism, at times, although when done well it straddles these worlds quite successfully) and you have four ways to describe the same thing.

The poor performance of historical technology predictions

As history shows, there are plenty of old trend reports that show the wrong trends. AI was predicted to become relevant in the 1970s. It has now gone through a ten-year hype cycle due to the breakthroughs in neural networks, but is arguably headed for another "winter" as the major progress in general AI has come to a halt (mostly because it has been ignored, awaiting better methods to tackle it, and because of high industrial demand for counting beans and analyzing images). Meanwhile, industry will continue to reap the benefits of machine learning progress that enhances analytics to streamline business operations. Robotics had at least three decades of constant hype followed by a silent decade before it suddenly exploded five years ago. The pandemic has made robotics turn another corner from exploration to implementation in factory manufacturing.

History shows us that the future cannot be predicted. I say this as someone who has spent a good part of the last few decades studying, acting on, and advising companies on emerging technologies and trends. Yet, serious futurists stopped predicting the future a long time ago. Instead, we advise around scenarios and paint pictures that allow executives to imagine what they might have to deal with and exactly how they would approach it, if current projections hold true. Besides, it is predicting *when* that is the real challenge, not predicting *what*, since the general "what" is typically known.

Why a framework of forces—and how do they work together?

It is one thing to say there is a multitude of forces impacting society and that technology is only one. It is an altogether different and more difficult matter to prove which forces matter the most. Even more difficult is to pick a technology and explain how it emerged and how it will change.

For our purposes, I focus on four main forces of disruption—technology, policy, business, social forces—and one additional "super force"—the environmental context that surrounds them. Other names for this approach are PEST framework (Hague, 2019) or PESTLE framework (CIPD, 2020), which at times include things like legal forces, demographics, or even ethics. For me, policy and regulation are highly interrelated, because there is no effective policy without legal review, implementation, and adjudication, so I consider them under the same umbrella. Demographics is among the social subforces and cannot be fruitfully understood otherwise. Ethics should be an overarching concern across the framework and not its own consideration because it gains legitimacy and force only from its spokespersons.

Another matter is that PESTLE is usually used to understand external challenges, yet internal organizational challenges could be even more detrimental, or, better yet, could be conducive to a deeper understanding and more headway. Therefore, in my version, each of the forces also must be used to understand internal organizational dynamics, without which you have not carried out an in-depth analysis at all.

The importance of conceptual rigor

The point is this. Deploying an empirically derived strategy framework is not just about which forces you choose, although the basic rationale for picking tech, politics, economics, social strata, and environmental footprint as relevant is not up for debate. What does need to be mentioned is that technology is not independent from each of the other ones and does not precede them, neither in time nor in function.

Instead, the real magic is in the conceptual framework that underpins each of them, how they interrelate, and how you can use it, bringing in empirical evidence. Many times, you need to carry out research to verify that what you think you have observed really matters in the analysis. Empirical realities on the ground also may change and evolve quite fast, so the observations must be up to date.

Putting mental models into a system

The popular blog and podcast I love to listen to, The Knowledge Project by Farnam Street's Shane Parrish (2020), obsesses over how mental models, defined as "a representation of how something works," are fundamental to capturing today's reality. Creating a latticework of some (somewhat) randomly gathered 109 mental models, Parrish attempts to prove the point that we need to have a hypothesis on how reality might behave (which is a good hunch), yet struggles to underpin each of them with sufficient evidence to prove that each is a useful contribution. The approach is inductive and iterative, and has initial appeal, but (still) has no internal structure and systemic theory of change underpinning it.

Part of the problem is that the word *mental* is not necessarily helpful in terms of what constitutes a model. What we should aim for is models of reality, which of course are little more than shared mental models by whichever community you trust to validate them. The danger of simply listing mental models absent a unifying conceptual framework is that it runs the risk of claiming that whatever observations you are making somehow are general and universally valid. It risks entrapping us in logical fallacies.

Whenever you apply any model—including a strategy framework such as forces of disruption—it is important to realize that it is a drastic simplification and is not valid in all cases. Even the way the subforces play out, which we will get to in the chapters on each of the disruptive forces, might change depending on circumstances.

To be clear, both *having* a systemic theory (which could be wrong) and *not having one* (which leaves you without a map) could be problematic. What is needed is an empirical mindset that constantly checks hunches and theories against facts and observations and allows for adjusting course without feeling down about it.

Without a stringent set of empirically derived observations about how each of the forces truly works, all you are doing is a high-school term paper on "various aspects," meshed together. This is, incidentally, why many people dismiss social science altogether. They assume that because they walk around in society and things "seem" so easy to interpret, the way things work necessarily also is quite easy to grasp. There is plenty of evidence from other fields that proves this is simply not true. What I will do in this book is show how specific and pinpointed your analysis must be to understand anything at all about how technology develops, changes, and impacts the world.

In consulting, mental models take on an even more purified form, often presented as a point of view (POV). The idea there, among consultants, is to show that they know what they are talking about and are not going in "green" when helping clients in a specific field. The challenge, again, is that every POV is necessarily myopic and might cloud as much as inform the client. This is why I, myself, don't typically use a POV in a collective sense, and do not recommend it to be used in business. POVs are highly personal and when they are shared (as in agreed among a set of informed participants), they can be extremely impressive and well thought out, and could lead to joint action. However, *before* these models are truly shared (and thoroughly discussed as such), they should not be represented as shared. Most consulting POVs fall into this category and have little value.

Simplify but do not assume total knowledge

Tracking *changes* in emerging technologies, policy, business, and social dynamics can help you determine which priorities to chase, which interrelationships to be aware of, and what gaps to fill. Focusing on changes is helpful in trying to decode the patterns that could give better visibility on future developments, but changes themselves are no guarantee. The speed, direction, and character of change could change at a moment's notice. Changes depend on scientific progress, products being formed, funding, and a myriad of other things, which I will cover throughout the book.

While simplistic, distilling the forces of disruption is a far better approach than chasing "all" the hottest technologies and tracking whatever information you can find on the internet. It is also a superstructure within which you can put your own mental models, or conceptual frameworks, or experiences, whatever the language you use, and ideally put them to the test.

In working with the most advanced organizations around the world, I have synthesized the learnings of many well-known strategy frameworks into one: the *Forces of Industry Disruption Environment Framework* (FIDEF), or simply "forces of disruption." You can use it to understand the future evolution of technology (which we do in this book), or to understand external forces impacting your organization (the way you would use a PESTLE framework), but you can equally well use it to understand your own organization or to compare two organizations (the way you might use a benchmarking framework). However, the subforces I delineate in the following chapters are specifically tailored to understanding technology.

Five spheres of future-oriented activity

This framework aligns five equally important factors impacting change in our time. Most importantly, this framework reveals how you might capture value from disruptive industry trends that are intertwined with technology developments.

Incidentally, understood this way, the topics are significantly narrower than simply describing the much broader spheres of activity, e.g. the technology sector (or the sum of all technologies around today), politics (in all its forms), the world of business (in a generic sense), or the economy (understood as the macroeconomic environment), or even any social phenomenon (the entire domain of sociology). That's where the difference from earlier frameworks of this sort becomes most apparent.

Each disruptive force points to which specific areas of activity to look for. Viewed that way, each of the forces is more than simply a convenient bundle of topics that allow you to keep track of a number of other related topics. As you see in Figure 1.1, these forces can be understood as simple headings, or I can add a little more specificity to each force. That way, the five forces are: (1) science and technology; (2) policy and regulation; (3) business

Figure 1.1 The socio-biological forces of industrial disruption

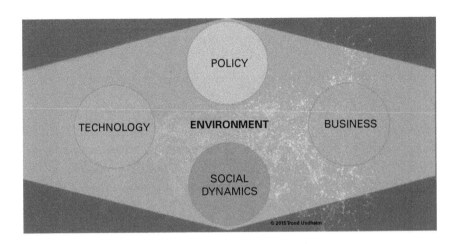

models and industry or marketplace concerns; (4) social and cultural dynamics, user concerns, and consumers; and (5) environmental footprint, e.g. places or processes where the physical environment directly intersects with human activity. The buckets are the same but there is slightly more color and specificity to each.

In fact, the simplest way to think of these five forces is as spheres of specific, ongoing activity: technology development, current or anticipated policy and regulatory activity, actively evolving business models and highly visible or underestimated aspects of social dynamics (typically understood as either social movements or consumer dynamics), and the expansion of cities or other built infrastructure into previously undisturbed nature as well as human-created pollution. That way, what to look for becomes very clear. Again, I need to point out that viewed this way, forces of disruption are much more specific than a brainstorming framework with buckets under which anything and everything might fit, which is typically how a PEST framework is used in case-based MBA education.

I will now go into some detail on each, being clear that the full analysis, with a few substantial examples of how each plays out, will follow in specific chapters.

Social and cultural dynamics

Let's start with social dynamics, partly because most people think of it as a reactionary force to technology. It is likely the other way around: first society prepares the ground for a scientific discovery by having certain challenges and being at a certain readiness, with a given availability of technologies, then new technology is invented as a response to a challenge or as a byproduct of doing something completely different.

Companies in the consumer markets are on to this. They no longer sit in internal R&D labs trying to reinvent their products. Rather, they come up with ideas, prototypes, and small line extensions and then test them with consumers in order to figure out whether these ideas have traction and how they need to be altered to be suitable for scaling up in a future market launch. There is a whole product development paradigm called Design Thinking (Brown, 2009), which builds on the importance of getting early user input before attempting to build anything of importance.

Business models and industry/marketplace concerns

Business models used to be quite commonplace: you bought and sold products on the marketplace, which often literally was a physical marketplace in the downtown center. Then, commerce moved into stores. Now commerce is taking place online, and the models have changed and bifurcated. The peer-to-peer markets are flourishing. People are liquidizing assets that for hundreds of years were viewed as relatively fixed and hard to buy and sell and impossible to rent or lend. Think of what Airbnb did to lodging—we are literally renting out rooms in our home to strangers. Airbnb was founded in 2008—it's strange to think that its business model doesn't sound strange anymore.

Another industry and marketplace concern is the changing viability of individual markets. While some traditional forms of commerce persist—the street markets in Marrakesh come to mind—other traditional forms are being reinvented. Consider the resurgence of farmers' markets, due to a renewed interest in fresh ingredients and farm-to-table proximity. Many new forms of commerce involve a return to ancient concepts of trust vested in familiarity, or in some form of proximity-based authority (the local merchant, the local farmer) which almost seems antiquated.

Policy and regulation

For any individual change, there are likely several surrounding factors that remain the same, which highlights the necessity of policy and regulation. When cars first arrived, there were no safety protocols and the roads soon became dangerous, with head-on collisions. As the realization that cars could kill emerged and the value of a human life increased, regulation caught up with line markings developed by the Road Commission in Wayne County, Michigan, in 1906, and evolving into dedicated lanes, lane width regulation, and further standardization.

Have lane markings slowed down automobile innovation? Perhaps, initially, because it demanded wider roads which were more expensive. However, generally standardization allows for a wider platform of innovation to get created, which enables more volume and scale.

The consequences of poor regulation are especially prevalent if a new technology fails, as voting technology has done numerous times. There is a backlash because of lack of public trust, and the result is that the evolution of voting technology has been painstakingly slow.

Policy and regulation are counterintuitive instruments of disruption. They disrupt initially because marketplace actors need to adapt. Tariffs, taxes, quotas, and duties are intrusive. However, as soon as that has occurred, and if relatively stable over time, regulation tends to level the playing field and enable fair, sustainable innovation based on a common platform.

At first, it may seem counterintuitive that industry would favor any kind of regulation. It is true that industry often resists for as long as possible. However, there are instances where regulation and standards become necessary and productive.

For example, in my previous role as a policy person for a technology company, I advocated a specific type of tech interoperability that would allow open standards to permeate software technologies across European public and private sectors. In that case, the software industry generally agreed that standardization as such was important, they just disagreed on how to deal with the royalties for the patents that at times surround these standards.

Not surprisingly, the companies that had built a platform and were extracting license revenue from patents (in that case Apple, Adobe, and Qualcomm) were keen to continue to do so, and those that saw more value in an open playing field (e.g. IBM, Oracle, and a host of smaller SMEs) felt differently. Today, as an Apple iPhone and iPad user, an avid content creator using Adobe's Creative Cloud, and a benefactor of many of Qualcomm's wireless technologies, I'm glad they have survived and continued to thrive even though their patent regimes are net negative for society as a whole. It's hard to predict whether a policy position will lead to desirable results over time. There is also, at times, a conflict between what's individually beneficial and what's socially desirable.

Science and technology

Why focus so much on science and technology, and not on art, or music? There are reasons why both art and music should be written up in a book about the future, but by sneaking tech into the title of my book, I want to stay on that topic as the main meme. Science is included because with so much short-term, incremental thinking around in ample supply, it is important to realize that most technological breakthroughs were prepared by science, and vice versa. A quick example to illustrate this point would be to consider the way the discovery of the science of DNA led to the technique of PCR, a technology that enables copying DNA, which in turn is helping lead to new scientific discoveries in biology and beyond (Understanding Science, 2020).

When you drill down into technology, there is much more work to do. I will expand on this more in Chapter 2. For now, in Figure 1.2 I've outlined a tree logic for the next level of categories that might show up in an analysis. On the technology side, we might want to consider subdividing technologies into whether they are legacy, infrastructure, mass market, or emerging technologies (and everything in between).

On the policy side, you should start the analysis with an understanding of what the current policy and regulatory environment is, then move to trends, risks, and stakeholder impact. You might end up with trends specific to each policy domain. Good policy makers always keep an eye on the stakeholder impact of their policies and run impact assessments to that effect in parallel

Figure 1.2 Subforces of disruption

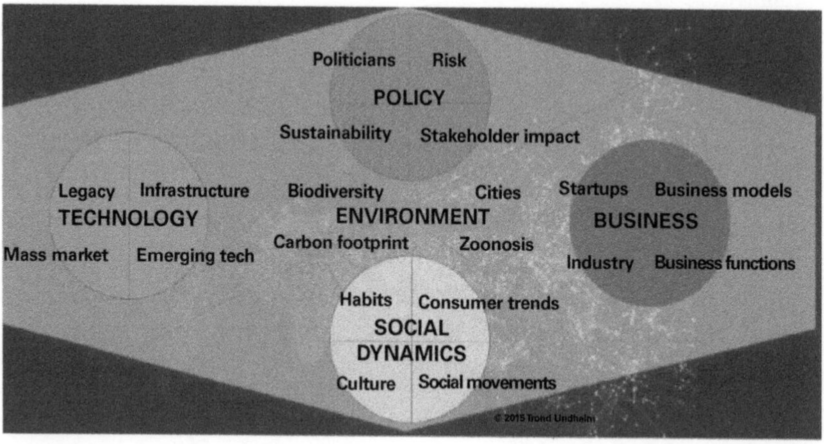

Politicians   Risk
POLICY
Sustainability   Stakeholder impact

Legacy   Infrastructure   Biodiversity   Cities   Startups   Business models
TECHNOLOGY   ENVIRONMENT   BUSINESS
Mass market   Emerging tech   Carbon footprint   Zoonosis   Industry   Business functions

Habits   Consumer trends
SOCIAL
DYNAMICS
Culture   Social movements

© 2015 Trond Undheim

to, or ideally before, finalizing policy. You must also consider the impact of lobbying as well as the political platform of the ruling politician(s) and party.

On the social side, generations might have a differential impact on issues and consumer trends, and beliefs vary between social groups and cultures. Social movements might form to try to reshape issues more to their ideal outcomes. However, the habits prevalent in dominant social groups are not likely to change as fast as most advertisers think they do, and it is worth spending a lot of energy understanding this in depth.

On the business side, quite distinct actors emerge: industry brands, start-ups, and VCs are distinct voices that impact disruption through products or funding. Meanwhile, business models might alter the positioning (or even existence) of any or all of these. Business functions will attempt to adapt correspondingly and will impact the search for talent and the types of training needed.

The environment

There are two types we should be concerned with: environmental footprint (ecology) and physical footprint (location), but most importantly it is the interaction between them that causes the challenge.

Whether environmental factors are explicit or implicit, disruption does impact the environment. Moreover, the environment is not a silent actor. The environment has various spokespersons in the form of environmental organizations (e.g. Greenpeace), policy makers (the "Green" parties), industries (the renewable industry), and even new technologies (sensors). The environmental force is perhaps the clearest example of all of the other four forces interacting, which may explain why it is so difficult to mobilize to save it. The complexity is somewhat overwhelming.

Infrastructure and location are two other concerns related to the place-based footprint that have to do with the physical and spatial concerns surrounding disruption. As the pioneering innovation studies by Dutch-American sociologist Saskia Sassen (1991) have shown, the reason New York still is the center of finance has a lot to do with the built-up infrastructure on Wall Street—the density of buildings and actors involved in that activity cannot easily be matched by another place (saskiasassen.com). Newer innovation hubs, such as the startup hubs around Silicon Valley and Boston, exhibit similar place-related advantages, which enable the kind of critical mass that takes years to build.

Disruption of demand

The demand for products or services is one instance where the four forces are quite useful. Whether you are trying to introduce a tech product or a fashion retail accessory, you are dealing with a mix of issues that are hard to pinpoint. Technology choices in the marketplace impact what it is possible for a new entrant in the market to do. Regulatory policy puts marketing restrictions, perhaps even tariffs, on certain products. Business model choices affect profitability. User dynamics might dictate a specific marketing approach or even completely tailored products for each relevant target buyer profile.

Each is a dynamically evolving set of constraints. On the path to profitability, disturbances may happen in one or each of these forces, and the way to address such challenges may entail acting with awareness of the whole demand cycle, which is likely not linear from one (say tech) to the other (say user dynamics), but perhaps the other way around, or careful interplay.

Given that the FIDEF framework is very intuitive, it can also be used to determine where you need to spend the bulk of your energy on a given tech trend or new development.

As we will see in further examples throughout this book, the balance will also shift over time. For instance, at the beginning of the 2020s, the entire software sector is in the early stages of a period with heavy regulation, which stems from its increased importance in people's lives. Other sectors with legacy technologies have been in periods of deregulation. Specific circumstances, for instance the appearance of creative, new business models (e.g. on demand), or a set of social movements gaining widespread traction (e.g. around wellness), can uproot this logic and create a more unwieldy mix of concerns.

Disruptive technologies are often related. Sometimes they depend on one another.

Visualizing relationships helps illustrate how by concentrating on a few concepts and causal paths at a time (even though reality is more complex). Reflecting on these connections enables learning. When learning is put together with contextual opportunity, we create insight.

In Figure 1.3, I suggest some ways that existing emerging technologies relate to each other. That kind of mapping could be done by anyone who tracks such technologies and undoubtedly generates new thoughts about the interrelationships. When discussed in a group setting, they can be very instructive as well.

Figure 1.3 Visualizing disruptive technologies

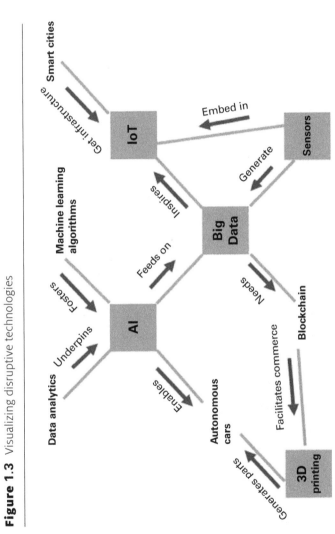

The social biosphere of innovation

There is another, even more important reason to think of interactions. At the heart, *forces of disruption* is more than a framework for understanding something. In fact, it is also a *socio-biological* theory of change. Within each of the forces, there are always living things and entities (humans, organizations) that play multiple roles (as scientists, politicians, executives, customers, advocates, and more).

In fact, there is value in thinking of the whole system of forces, and each force as well, as living things. If you think of the four main forces together as an organism or even as constituting a "biosphere," akin to the thin surface layer that supports life on Earth, it becomes more evident how you cannot ignore any parts of the system, even if a part appears dormant from one moment to another.

The benefit of having a systems theory

Let's assume the world can be understood in systemic fashion (I sure hope so). So, what does it mean that the world is a set of systems? There is a whole field of engineering called system dynamics (Forrester, 1961), which can be applied to any problem characterized by interdependence, mutual interaction, information feedback, and which nowadays can be greatly enhanced by both visual and computer modeling efforts.

There is also a whole field of emerging social theory called social physics, developed by extreme functionalist Sandy Pentland (2015) from MIT Media Lab, which studies patterns of information exchange in a social network without any knowledge of the actual content of the information and can still predict with accuracy, greatly enhanced by contemporary digital monitoring technology, such as machine learning-enabled cameras and sensors.

A myriad of other systems-based theories also exists, simultaneously trying to structure this emerging world. While we do not necessarily need to know the ins and outs of such theories to understand the emerging world, we do need to have an inkling. What those theories often have in common is a disdain for the effort that goes into understanding human intention. However, it remains true that simultaneously as it could be understood in systemic fashion without human input, so to speak, individuals, regardless of the multiple roles they play, contribute to the mix as socially situated beings, in a specific context, with a history of interactions, a trajectory, and a set of future projects. In either case, what disruptive forces or subforces are *not* is mechanistic "tools" that have only clear-cut, standardized, and controllable engineering functions associated with them.

Somebody always speaks on behalf of technologies

Even the technologies themselves cannot be fruitfully understood as passive "tools," because to be effective, they need human representation. Think about it. No organization would invest in technology without an executive arguing it should be done.

What I mean by that, following French philosopher-sociologist Bruno Latour, is that whether it be AI, robots, 3D printing, IoT, or sensors, perhaps particularly evident with AI, these are not (yet) actors in their own right. AI does not yet have its own voice; it is all human-supported.

First off, for AI to apply, humans have to argue that there is room for AI in the first place. Then, somebody has to acquire the resources to put it into action, pay for the computer, create the software, insert the data, tune the model, study the results, and apply the lessons. What's more, all of current AI is extremely domain-specific. It can do a given thing only within a highly specific and specified context. It can reproduce answers, which humans see as desirable. When computers, in rare cases, do not, or we do not understand their analysis, we typically shut them off—or we should. We cannot afford to have machines doing what we don't understand.

Robots are no different—they are created by humans with specific actions in mind. Humans then put together their hardware, install their sensors, and ensure the software works. Humans also have to put the robots in a context they can master and must control that things go well and shut them down when they don't. The fact that there are so many human steps involved even in automation is also a safeguard. What it means is that within "narrow AI", there is a very small chance that AI will run awry. Quite the opposite, current AI still has a mind-blowingly narrow application space.

Which technology do you speak for?

One way these kinds of issues come into focus is when we, as one does, start to identify "issue champions" for various emerging topics that become relevant in an organization but don't yet have a clear owner and path forward. These issue champions might have created the issue in the first place or might just be vocal advocates of a specific approach to the issue. One friend told me that it took him only a few months (this was about a decade ago) to become "McKinsey's go-to person for health-care tech," which was an emerging area at the time but now obviously is a big line of business in the firm.

The point is, you can likely quite easily attach just a few names to each technology, and you'll know who to go to, whether inside your company or an outside expert. In fact, if you have trouble doing so, you likely have chosen the wrong technology focus, or you might truly have a challenge in that domain and need to act.

As we engage with the future of technology, we should make ever more conscious choices of which technologies we want to get involved with, spend time understanding, and, quite frankly, allow be part of our sphere of concern. We cannot make these choices entirely independently of what is going on in the general biosphere around us. We also cannot simply adapt what we think are givens from a trend report we just read, or from the shorthand of a company's strategy documents. However, our attention, as individuals, organizations, countries, or whatever type of social group you consider, is limited. We need to use it wisely.

How many technologies do you need to master?

As will become apparent in Chapter 2, engaging with 3–5 technologies might be what's required of you at work or in your next startup project. However, that engagement will not be as flexible as the annual tech reports you read. Gaining basic awareness of a technology may take you only a few weeks. Mastering it may take you months or even years, depending on the strategies you use and the resources you have available to you.

At the same time, you need to be aware that your own focus does not necessarily change what others are focused on, so while you might have one focus, the overall system might move in a different direction.

The internet is perhaps the biggest example of a technology that is 100 percent kept alive by humans. Without humans sustaining and building out content, the network of nodes would have little meaning, and would also have little value. It is not the internet as a technology that has value; it is the internet as a social construction where a huge amount of actions keep it maintained and grow it further, hour by hour. The cables that sustain it, the networks, even the standards that underpin the internet were all negotiated by humans, paid for by humans, and are sustained by humans through continued investment and discussion.

Figure 1.4 Industry sector disruption grid

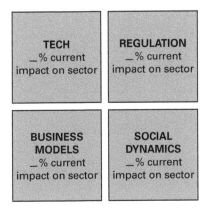

How do technologies layer on top of each other?

In the following chapters we will drill down on each of the disruption forces to figure out how each of the forces works and changes, and how they layer on top of each other. More importantly, we will try to establish patterns in how tech trends evolve. That way, you can maintain critical distance the next time you are presented with "this year's most important tech trends."

It is quite possible, for instance, that you should rather be picking from the last three years' tech trends (because you have some critical distance), instead of focusing so much on this year's trends (which are highly speculative).

This is what *Future Tech* is all about, taking a small step backwards in order to (potentially) make a giant leap forward. Now, prepare to step back and consider the impact of science and technology.

Key takeaways and reflections

1 What tech trends are you aware of at the moment? Try to think of how you came to form this opinion. Which reports have you read? Which news sources informed you? Is there a specific individual, startup, or company associated with these trends?

2 The four forces of disruption is obviously a simplification. What are some other candidates that also are relevant "forces"?

3 Try to run a quick analysis of Four Forces of Disruption for the following: the local supermarket (if it still exists), the World Bank, and the Olympic Games. What are the forces in common influencing all of them?

4 Consider that it might be possible to evaluate the percentagewise distribution of disruption emphasis for each industry sector, technology, or whatever phenomenon you are looking at. Use the industry sector disruption grid (see Figure 1.4) to go through a few areas you are concerned with and fill in whether technology, regulation, business models, or social dynamics are currently impacting that area the most. You can also create a few boxes and try to map this on a timeline of 1–3–5 and up to 10 years ahead. You can bet that savvy investors or regulators are doing something similar (if they are as smart as you). After some initial analysis, by assigning percentages to each quadrant, you can properly allocate resources. These percentages could be determined by your own team or you can use external metrics to try to assess what the right balance is.

How science and technology enable innovation

02

In this chapter, I look at how to understand science and technology through three lenses: platform business, taxonomy, and tech visualization. Platform technologies are the tools that make things happen and tie in with business models. Taxonomies are the concepts that organize a field and are the basis for true domain understanding and communication ability. Tech visualization is a means to quickly grasp complex relationships and can act both as a shorthand and an accelerator in making connections between fields. Either way, I will show that deeper engagement with sci-tech will be needed from all who want to understand or profit from the changes coming in the next few decades. However, you need to deeply consider crosscutting issues, not just domain skills. Similarly, technology is also reactive, in that it is often developed as a response to a call for action, a request, government funding schemes, or business goals within corporations. At times, technologies are part of long chains of actions and fall into a pattern of evolution which other parts of society start to depend on, such as vaccine development or regular computer speed upgrades.

Tech discovery approach #1: grasp how technology platform businesses work

Startups valued above $1 billion, following venture capitalist Aileen Lee's (2013) popular terminology, are called unicorns. According to CB Insights, there are near 500 unicorns, up from 200+ in 2017 and only 80 in 2015.

Between 60 percent and 70 percent of all startup unicorns in 2017 were platform businesses (Cusumano, 2019), which means that they create value by facilitating exchanges between two or more interdependent groups, usually consumers and producers. The metaphor of "platform," which of course originally referred to a raised level surface on which people or things can stand, has become quite popular over the past decade. These businesses create a surface which other businesses can use, and indeed in many instances are forced to use in order to have a business at all.

The main types of platform technologies in operation today facilitate transactions, exchange, interaction, making and sharing things or content. The best way to illustrate that is to look at small, third-party sellers on Amazon, the world's leading online retailer. Amazon facilitates efficient e-commerce transactions both directly and by connecting buyers to third-party vendors. Without Amazon, these vendors would have no viable business model since they would not otherwise be able to reach a worldwide market. Alone, they are simply small vendors of niche products. Together with Amazon, they provide a distributed product network that, in sum, has great benefits for both parties (although mostly for Amazon).

Kraken, the cryptocurrency trading platform, facilitates *exchange*. Fiverr, the online marketplace, facilitates selling creative digital services. I have used it many times, and the approach is quite brilliant both for Fiverr, which takes a 5-percent processing fee, for the sellers, which now have a worldwide market for their digital services and can try to get clients (starting at $5 but rapidly ascending once you start to rely on their services), and for the clients (who enjoy inexpensive digital services with a basic quality assurance, although what that means can depend quite a bit).

Udemy, the e-learning platform, facilitates creating and offering online courses, which otherwise also suffer from the same challenge: how to find a market for a course and how to gain access to a great diversity of online courses.

Uber, the ride-sharing app, facilitates *interaction* by lowering transaction costs. The service has also turned out to be useful for food delivery because the network of drivers is now so vast, covering enormous territory, which is convenient for restaurants and grocery stores using them as a delivery service. Why did Webvan, founded in 1996, fail (in 2001) but Uber seems to thrive? The demand curves are radically different. Even at its peak, Webvan existed in only 10 US markets and not abroad. Webvan also built its own fulfilment centers and took on parts of the value chain that were too complicated for being so early in the development of online delivery. Hindsight is, of course, 20/20.

The video-sharing platform YouTube enables *making* and *sharing* things or content created by yourself or by others. Most are recent phenomena enabled by the web as the proto platform underlying each specific platform, and the majority depend on some form of digital technology.

Figure 2.1 shows the various platform technologies.

The relationship between emerging tech and platforms increases probability of success

Emerging technologies, in fact, will typically be introduced in earnest only once there is a set of successful platform technologies (and often a set of built-in users) to accompany them. This was the case with computers that depended on the TCP/IP email protocol and a network of computer owners, who had a generous benefactor in governments and universities.

Gene therapy had the backing of the pathbreaking CRISPR approach as well as cohorts of sick patients willing to try everything. Internet of Things became a real thing only through 4G and 5G telecommunications networks providing the speed to allow real-time edge device communication, as well as an install base of smartphones and digital devices that can connect to it. Robotics have become highly successful of late because of a fortunate triad of IoT, batteries, and inexpensive highly functional sensors, plus, lately, the lack of competition from humans who have resorted to working from home as much as possible due to the COVID-19 pandemic.

Figure 2.1 Evolving platform technologies

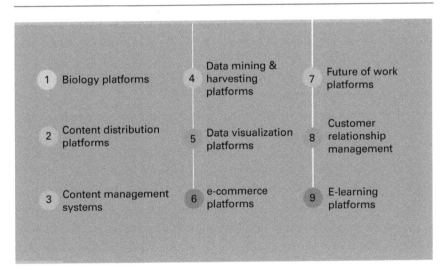

Finally, digital currency has benefited (indeed become possible) because of the blockchain protocol, because of banks willing to experiment, because (many) governments have not shut them down but have been patient to see what would happen, fueled by consumers unhappy with bank fees as well as by millennial investors keen to experiment.

However, as we have seen above, in addition to relying on other technologies, emerging tech needs an attractive use case (or two) and favorable consumer dynamics.

Artificial intelligence is an example of an emerging technology which is still looking for a precise game-changing use case, although its current form, neural nets, are quite the hit in government surveillance, operational analytics in industry, and financial modeling. Weather analysis, image recognition, and chatbots were early narrow examples of such use cases, but they will not be enough to propel it forward into a mainstream technology, at least not without massive improvements to the form factor of its delivery.

The COVID-19 pandemic is an example of an event that creates higher demand for videoconferencing, and longer term creates real demand for virtual reality, another technology previously without a compelling use case beyond gaming and architecture.

Tech discovery approach #2: understand how taxonomies work

Taxonomy is the science of naming, defining, and classifying things. It originated in biology where the task was to organize the variety found in nature so biologists could speak about their discoveries more easily. More generally, taxonomies are ways to organize knowledge in a standardized way. Taxonomy is useful far beyond biology. The simplest taxonomy may contain just three terms, but if those three terms are exhaustive, this can be highly efficient and insightful. The simpler the taxonomy, the better, although they tend to form tree-like structures that relate concepts hierarchically and can get messy both to illustrate and to remember.

The issue of visualization becomes especially pertinent in the age of machine learning. Many different techniques, statistics based or not, exist to display information. If you let loose one or several of these methods, you better know what you have created. Being able to interpret what's shown is the prerequisite for showing it to others. This should become an ethical norm. Emerging thinking in the ethics of AI emphasizes explainability as a key goal.

Classifications act like filters that allow us to control the inflow and out-flow of ideas and facts. When you read a stream of blogs or mainstream media articles about technology, there is no organizing principle at play, ex-cept the journalist's or blogger's claim to be writing about something relevant and emergent. As human beings overwhelmed by impressions, we need or-ganizing principles. Without them, we are just empty vessels that fill up with whatever comes our way until we are saturated and cannot digest more.

Classifications are the beginning of organizing our knowledge into bite-size chunks. Classifications are especially helpful in understanding how sci-ence and technology emerge. They are also a useful lens for understanding the development and growth of a field, as specific classifications get added or abandoned over time.

Sci-tech taxonomies start with scientific disciplines, continue through tech trends (including startup application domains such as Edtech and Fintech), encompass use case examples such as B2B and B2C applications as well as global challenges.

Basic classifications in science

Here's a comparably simple example regarding the branches of science. The main branches of science are physical science, earth science, life science, human science, and social science. At face value, this distinction would seem simple enough, even mundane. However, we have already made a controver-sial choice of including two historically "softer" sciences (humanistic and social science) into the classical triadic distinction which did not include them. Moreover, the classification obscures the overlap and collaboration between them through multidisciplinary collaboration. This illustrates how even simple classifications can either add value (because of their clarity) or at times need to evolve as domains develop and connections start to form across domains. The "branches of science" taxonomy is clearly due for a makeover and my educated guess is that it will happen over the next decade.

For example, I would remove the distinction between physical and earth sciences (not sure why it exists). I would also argue that social science is humanistic (or vice versa), and that computer science should be its own scientific discipline but perhaps it should be called digital science or some-thing more fancy such as connection science. There would ideally also be an innovation science, but I happen to believe that innovation is so complex that it defies scientific description.

The book you are reading now might be as close as you will get to putting tech and innovation into a structure. The problem is that the moment you do, you will immediately be superseded by the complex and evolving reality around the corner. No one problem is the same. Methods continuously evolve. Capabilities increase. What counted as scientific method a year ago might be outdated the next. Conversely, some principles stay the same for 100 years. It is just difficult to know which. If you fully knew, you would be kingmaker in science and technology funding, university organization, and ultimately technological progress.

What does this simple classification into branches of sciences do?

First, naming something matters and has a variety of nuanced, highly specific, and evolving cultural, political, and scientific connotations, sometimes hidden to the uninitiated. Incorrectly naming something or calling it by the wrong name or classifying it in the wrong way is misleading. If you care deeply about a topic, you need to have the appropriate categories within which to talk about it with others who care and know what they are talking about. Only then can you move on to studying behaviors and more complex things than names. Whether it's mountain lions, computer games, musical instruments, or home improvement, the point is, you need to dig deep in order to extract true, valuable insights. Staying on the surface will only make you a consumer of knowledge, not a producer.

Second, when it comes to innovation, it is the spaces in between that are important, which is not to say that classification is not possible, needed, or useful, just that it may not be sufficient. I would argue that in this case, the "five branches of science" mnemonic is a highly generic tool that clouds the issue of the interrelationships. The fact that they used to be wildly separate paths in university is, in my opinion, part of the reason why scientific progress (and industrial application) has not moved faster than it has done over the past few hundred years. That being said, each of these five branches of science could be a starting point for further investigation.

The strange relationship between science and technology

One of the reasons why taxonomies matter is because they artificially structure what is inherently messy. Even the very basic distinction between science and technology is. Science is related to basic learning. Technology is related to application. That's the theory, at least, and it's what is important in a regulatory context. However, even if science and technology foster innovation each in different ways, there is no linear path from science to engineering via technology to application and products, although it may look like that at face value. Rather, the process is more like a series of loops where they depend on each other. The investment term *sci-tech* a shorthand which I have adopted in this book for convenience, looks at the joint impact of science and technology, but equating the two is also quite misleading.

An intrinsically better way to see the two is to think of science as the knowledge system (in place since the 14th century) and technology as the technique used to control something (formally adapted much later, in 1859), including scientific elements (Li, 2003). The confusion between "sci" and "tech" in my view also stems from a lack of consideration for exactly how science and technology, each in different ways, contribute to innovation. The process is not linear, as I said above, but rather full of starts and stops. By comparison, science is messier than technology, which I will turn to now.

Science is disorderly, technology is orderly

As Latour and Woolgar (1986) showed in their study of how scientists work, an inordinate amount of energy is spent "producing ordered and plausible accounts out of a mass of disordered observations." This observation is still shocking to many of us because we often assume that science is fixed in facts and realities. It is, but these facts are created by consensus, not by some mysterious access to nature itself. Although we constantly create theories and descriptions that (more or less) make sense to the scientific community, we simply don't know anything about nature itself. Nature is outside the realm of science. This fact, by the way, complicates even "first principles" thinking, because there are no first principles that are immune to this logic, e.g. in some ways there are no first principles accessible to humans at the moment.

Technology is quite different. That's when it becomes apparent that technology (typically) appears much later in the timeline. By the time something is ready to make into a heuristic, a set of instructions, or perhaps even to create a platform around (e.g. our previous discussion of technology platforms), there is already a user base, a set of professionals who understand more or less the same by the key terms and approaches used in the domain, and also a somewhat clear roadmap of what needs to be done next.

It is important to maintain this distinction because otherwise, by inadvertently making the disorderly too orderly, scientists might obfuscate the complexity involved in creating order. And because replication in science is exceedingly rare, there is little verification of just how complex the process of discovery is. It is in human nature to simplify things.

This book is an attempt to simplify the understanding of the future of technology. This chapter, by itself, is a way to simplify the description of the activities that are at the core of technology proper (as opposed to the wider aspects, which we spend considerably more time on because they are in fact more important in the total picture).

In doing so, we focus not so much on the discoveries themselves (because they will evolve) but more on the people that engage in discovery, and the institutions and phenomena that are changed or come into a different light because of tech. Furthermore, I want to get at the curiosity, not just the results. The best scientists are sometimes great educators, too, but not always.

The path from science to tech is not linear—there is also a path from tech to science

There is a strong tendency to consider science and technology as one thing—even though they are quite distinct. It's also not as easy as to say that basic science precedes technology. Often it is the other way around, where a new technology leads to scientific breakthroughs. For instance, Understanding Science (2020) point outs that the X-ray machine led to the possibility of X-ray crystallography, which enabled scientists to see snapshots of molecular structures, which ultimately led to the discovery of the double helix atomic structure of DNA.

Finally, science sometimes emerges from societal demand. For instance, vaccines are rarely commercially successful. The only reason they happen is because of the public health prerogative, government intervention, funding or legal mandate (or all of the above), as well as a sense of heartfelt urgency among a subgroup of scientists. In the five years from 2020–2025 this is happening surrounding the race to develop a COVID-19 vaccine. In that case, it may even give meaning to say that the science stems from a societal need and didn't originate in the realm of science at all. We will get into further examples of each of these cases in a moment.

Classification of technology within society

One categorization that I find useful is focused on the degree to which the technology is embedded in society. Why is this classification useful? Social construction of technology (SCOT), of which tradition I am a relatively reluctant disciple, argues that technology does not determine human action, but that human action shapes technology (Bijker et al., 1987). In other words, technology cannot be understood without understanding how that technology is embedded in its social context.

Consider the social construction behind one of our most invaluable technologies—the internet. The internet started out as an experimental way to communicate between scientists. It has evolved to become crucial for the ways in which goods are discovered and sold, how governments communicate and deliver services, how we work, as well as becoming essential to how children and youth play and learn. Each of these use cases has evolved through a complex process. There was nothing set in stone about how the internet would evolve based on its original design. And it continues to evolve as more social groups overcome the hurdles to access and develop the technical skills to maximize its impact.

Classifying technologies in terms of how they are used in society is not simply a case of an adoption curve, in the vein of what Gartner's Hype Cycle (2020) attempts to do. It begins with dividing technologies into how they are embedded in society, which I see as breaking down into the following categories: emerging, infrastructure, common, mass market, and legacy, listed after degree of future orientation, as well as with some examples. This is shown in Figure 2.2.

Figure 2.2 Deep dive into tech disruption forces

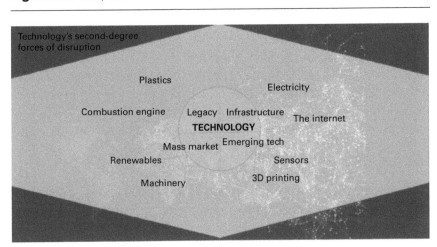

At the next level, I would attempt to create a sublevel cognitive structure which, at minimum, had a few examples of each and ideally indicated 2–3 subtypes. For example, frontier tech includes sensors and 3D printing, infrastructure tech includes electricity and the internet, and legacy tech includes plastics. Using that logic, one might be able to discern a lot about the timeline and future evolution of these technologies, and perhaps even their place in the forces of disruption chart overall. However, as I'll demonstrate next, there is not always agreement on where a technology needs to be classified, or indeed it could depend on geography and access, which is related to social structure and inequality.

Emerging tech

Emerging tech (which relates to other concepts such as frontier tech, deep tech, hard tech, sci-fi tech, exponential tech, or tough tech) is technology that still has an adoption curve into the future. It is aspirational, complicated, unproven. It may have high potential, but it is not yet possible to scale its production. One example of current emerging tech is 3D printing. Although we now have desktop 3D printers, the true potential of metal 3D printing, for example, is still only possible to see using prototypes and early models.

Having said that, Desktop Metal, an MIT startup founded in 2015, recently put out two metal 3D printing systems covering the full product life cycle, from prototyping to mass production, and began shipping its product in 2019. The potential of such additive manufacturing approaches is thought to increase only once the systems can scale to larger build spaces and will enable far higher production speeds (Kritzinger et al., 2018).

Because of availability, skill level required to deploy or make use of it, or even just price, emerging tech is not adopted evenly across the globe. One nation's frontier tech is another nation's infrastructure tech. While tech firms such as Gartner or Forrester may cloud this issue by assuming linear adoption, linearity is simply not the case. The highly influential framework called diffusion of innovation (Rogers, 1962) is so misguided, yet still so pervasive in consulting circles.

Infrastructure tech

Once a technology has become part of the fabric of society, it becomes part of its infrastructure. Infrastructure technologies, such as electricity, occupy such a central place in our society that we don't need to understand their inner workings in order to use them, although it is still quite helpful to know before you attempt changing a circuit breaker. Similarly, while it's useful to know what a TCP/IP standard means, it's no longer required knowledge for using the internet. The crucial aspect of these technologies is that they represent either a physical infrastructure or a combination of a hardware/ software solution that locks us into certain paths that prove costly to get around. Very few technologies get to the point that they are considered essential and underpin nearly every aspect of our life.

Common tech

Common tech are those technologies that achieve a certain level of adoption across society yet don't quite represent platforms on their own. Consider computers or podcasts. The computer is highly fundamental but isn't what creates the shared common good. The podcast is an interesting application but depends on a whole set of surrounding factors (uniqueness of concept, attractiveness of content, content quality and quantity, audio, fame, digital savvy, marketing budget, word of mouth, Really Simple Syndication (RSS)) in order to become efficiently spread.

Common tech are common technologies at various stages in our society's development, but they are likely to change without requiring all of society to rewire anything to allow for that change to happen.

Mass market tech

Once common tech reaches mass consumption, they change character once again and become mass market tech. Mechanical machinery such as electromotors or fuse boxes are widely available and power many technological devices on the market today. At the same time, mass market tech also tends to consist of commodity components that are replaceable, which is good for serviceability but bad for individual businesses attempting to lock in clients.

Legacy tech

In software systems, legacy tech refers to older systems that typically have poor interoperability—they cannot communicate easily with newer systems. Legacy technologies have been around for a while but are in danger of being replaced by newer, better alternatives. Plastics, as useful as they are, will (hopefully) gradually give way to organically created materials that are more sustainable. Electricity is now not the only way to supply energy to a home or to a light source.

Tech discovery approach #3: visualize the relationship between technologies

Disruptive technologies are often related and might be dependent on one another. When trying to understand how technologies are related, start by concentrating on a few concepts and causal paths at a time. This enables you to isolate phenomena enough to study them and reflect on the connections you see between them.

When you understand the relationships between technologies, you can look for opportunities to apply that insight. This insight becomes a potential innovation that you can act upon. Next you would want to produce a prototype, test that prototype, get feedback, and potentially go into production and sell something based on the initial embryonic idea we had long ago, much earlier in the thought cycle.

Visualization methods

The issue of visualization becomes especially pertinent in the age of machine learning. Many different techniques, statistics based or not, exist to display information. If you let loose one or several of these methods, you better know what you have created. Being able to interpret what's shown is the prerequisite for showing it to others. This should become an ethical norm. Emerging thinking in the ethics of AI emphasizes explainability as a key goal. Tag clouds became popular about a decade ago for their ability to present a complex array of concepts in a quick way based on frequency of the key terms (see Figure 2.3). The utility of the tag cloud depends on the care you took to create it. Mere frequency of terms is not sufficient to make a judgement about what a text is about or how to organize your thinking about a domain. Simple visuals like tag clouds should be used to illustrate only very obvious things.

The issue of whether you have the skills required to read a tech visualization is one thing. Being able to create it is another. Creating advanced visuals used to be complicated but is becoming easier and cheaper. Data visualization tools including Google Charts, Tableau, Grafana, Chartist.js,

Figure 2.3 Technology tag cloud

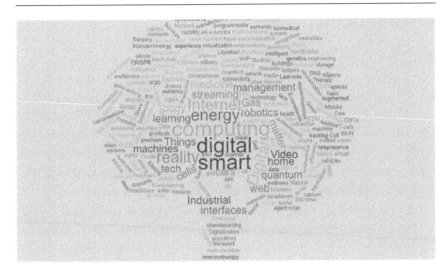

SOURCE https://www.wordclouds.com/

FusionCharts, Datawrapper, Infogram, ChartBlocks, and D3.js each offer a plethora of options. In fact, the best tools offer a variety of visualization styles, are easy to use, and can handle large data sets. Some of these are free, such as D3.js., ggplot2 (written in R) or Matplotlib, Seaborn or Bokeh (written in Python) programming language, or offer freemium versions. Many of them enable businesspeople to explore on their own, although some tools take more tech skill than others. For the purposes of using forces of disruption analysis, even simpler visual tools, including Microsoft PowerPoint and its equivalents as well as the new graphic design platform Canva, can be quite useful.

Visualization is helpful not only to communicate insight to others but also in order to structure your own thinking. If you cannot illustrate your insight in a quick visual, that might be a sign that you don't really understand it deeply enough. As visual software becomes easier to use, there is no excuse and experimenting with Canva can be a learning tool in and of itself. In many ways, these software tools impact our thinking, the way that PowerPoint did for an earlier generation a decade ago.

The figures used in this book are also examples of visualizations that, if successful, yield a concentrated picture of reality without necessarily going via the much slower comparably repetitive medium of written text.

Tech discovery approach #4: track specific technologies

The forces of disruption framework can also be used to look at individual technologies. In fact, this may be an instance where the framework is particularly useful. The technologies that matter to an organization will change depending on the challenges the organization faces. Whether the technology is of great interest in terms of tech scouting will obviously depend on whether the internal capacity in the domain is sufficient or not. However, the opposite is also true. If an organization is specialized in a technology, it may have a special interest in discovering what external parties are up to in that domain, either to learn, to share, to collaborate on, or to validate existing approaches. With that in mind, let's create an interim taxonomy for some technologies that have been reasonably popular to discuss over the past five years.

We chose the top 15 technologies in order to cover a wide enough variety of technologies to cast a wider net of relevance to as many readers as possible (see Figure 2.4). We are not going to go into each in depth but rather will choose a few of them. Executives typically don't focus on more than 3–5 technologies at a time.

The challenge with this particular taxonomy is that it has a quite random structure. The organizing principle is something like "terms that Trond found to be useful and popular in and around MIT, among both scientists and industry, in the period from 2016–2019." This does not make it irrelevant, but it fixes it in time and context to something that, inevitably, will change. Such an approach is not unusual, and it can have some value, but it is casuistic and not a reliable way to classify things in the long term. But this approach is popular with consultants and market researchers writing trend reports. They randomly search around for "signal" that a particular technology is "on the uptick" in terms of attention and simply declare that it is without a clear heuristic. Also, they are constrained by the fact that this year's report needs to look different from last year's report (yet still somehow consistent).

For a discussion of the five technologies I chose to feature in this book, see Chapter 6.

How to track tech trends?

Consulting firms typically publish annual reports on tech trends based on such lists claiming they are essential. See, for example, the tech trends from

Figure 2.4 Top 15 tech taxonomy

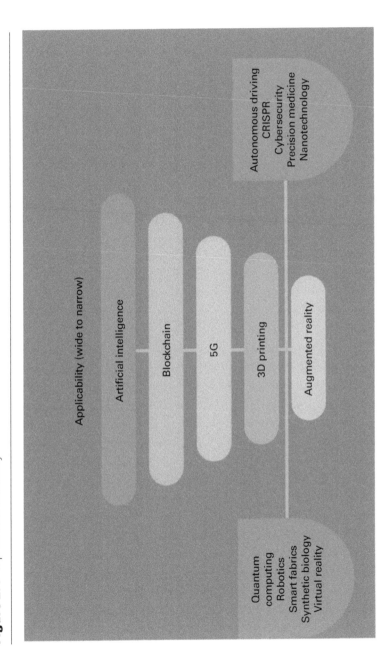

Gartner, Accenture, KPMG, McKinsey, and others. But why are such lists and similar taxonomies relevant? They are highly useful as necessary short-hand simplifications of the constant flow of rapidly changing technologies and developments.

In fact, in today's fast-moving society, shortlists are convenient time-savers. However, they are not necessarily the best to foster critical thinking. That does not mean I'm not recommending reading such lists, they just cannot be used in isolation but need to be one among many, many inputs.

Only a very few disruption experts have any capacity to track more than a few of these topics in some depth at any given time, which is why making use of a man/machine system to accelerate your learning and to track tech insight is going to become essential in the next decade. I often deploy the man/machine systems to select content from a shortlist of publishers I have already decided on. Then, as a second step, I plug the terms I find into one of the forces of disruption 4x4 matrices.

In this book, I argue that you should be extremely critical of anything you hear or read around technology. Ideally, you should attempt to create your *own* priority list (like the one I made on the preceding pages) and work from there instead, or as a second step in order to digest the various lists you read every month.

The problem with such lists created by others is that they often don't provide full background on how they emerged. They are typically not referenced like a science article. They might very well be written by authors who have little credibility in the field they are writing about, although sometimes that's not the case, it just depends. Were they really the results of deep research? How long will they be valid and in what context? Some of these reports are the result of months-long study, others were quickly put together in journalistic fashion. If there is a methodology section, you know at least some effort was spent in planning and research.

Such a random tech taxonomy might be entirely appropriate to discuss what emerging topics are top of mind in the marketplace. But does that mean much for what you should focus on? It does lead to a focus on similar things, which may lead to mediocrity. However, given the open-ended, complex nature of technologies, just having the label doesn't make you the expert. Once you know what topic to focus on, that's when the real work starts.

Let's illustrate this with an example where we drill just one or two levels down for a few technologies. Let's start with a general overview of what a technology is all about, who the players are, what the key events are, and what the policy context is. In Figure 2.5, I have charted what you might consider a baseline analysis of a specific technology.

Figure 2.5 Technology disruption subforces overview

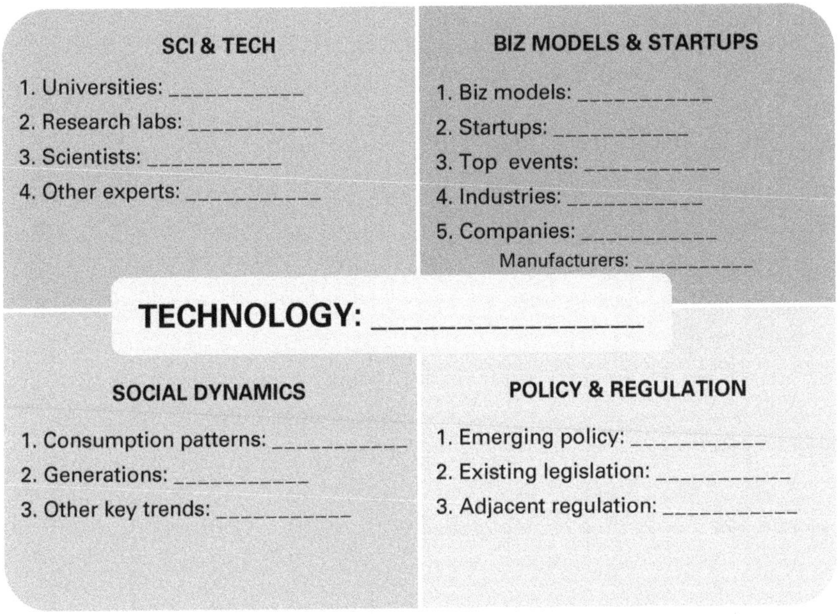

SCI & TECH	BIZ MODELS & STARTUPS
1. Universities: _____	1. Biz models: _____
2. Research labs: _____	2. Startups: _____
3. Scientists: _____	3. Top events: _____
4. Other experts: _____	4. Industries: _____
	5. Companies: _____
	Manufacturers: _____

TECHNOLOGY: _____

SOCIAL DYNAMICS	POLICY & REGULATION
1. Consumption patterns: _____	1. Emerging policy: _____
2. Generations: _____	2. Existing legislation: _____
3. Other key trends: _____	3. Adjacent regulation: _____

Each disruptive force has 3–5 subcategories. For example, under Sci & Tech, I'd suggest tracking a few specific universities, research labs, scientists, and other experts (influencers), and I'd include references to a few important scientific papers as well as summary articles. As a result of all of the above, there should be a list of 3–5 research paradigms or research methods that summarize the ongoing activity.

Under Policy & Regulation, I'd suggest tracking both emerging policy and existing and adjacent regulation (because of spillover effects). Under Social Dynamics, I'd focus on three things: consumption patterns, generations, and social movements. Under Biz Models & Startups, I'd list a few key emerging business models, a few relevant startups, a few relevant industries, and finally some larger companies that operate actively in the space.

A highly simplified version of Figure 2.5, having done an analysis of what matters, might look something like Figure 2.6. Note that there may not be a 1:1 answer to each question in Figure 2.5. Depending on what you are using the analysis for, this could be fine, or it might be an essential element of what you are trying to accomplish. The heuristic of using 4×4 schematics is not very well designed for scientific research where you need the full references to appear; it is meant more for overview and sharing.

How deep to dig?

Every field has its specialty areas. Depending on your purpose, you can stay at the surface or dig deep. Let's take computer science as an example. Artificial intelligence is part of computer science. That's perhaps an obvious fact, but also not completely accurate. AI is also part of psychology, of linguistics, in fact there is perhaps not a field that will not want to claim a piece of AI. But what about computer science itself? What does it consist of? Figure 2.6 gives a quite recognized taxonomy that represents my distillation of a common core of specialty areas that appear in the top university departments. It is also checked against terms used in top industry publications.

Without further qualification, such a list of topics doesn't tell us much, beyond saying something about what courses to expect in a college course in computer science and revealing the main (historical) research areas. The next level of insight might be gleaned from citing an article describing the history of each field, or the founder(s) of the field. I will refrain from doing so here, but you get the point if you want to explore. The list is definitely of higher validity than a market research report on trends in computer science. It has stood the test of time. But is it usable for a non-expert?

The reason I spend time on this topic is to illustrate how you can go about truly understanding a field of technology, starting with its academic origins and diving into subfields based on your current abilities. For each of these fields there are reference articles available on the internet that do not demand a lot of prior knowledge. There are obviously specialist sources, too, so you have to choose wisely.

How to contextualize insight

In the age of search engines you could always use it as search terms. There are also online courses on most of these topics. But in order to be truly useful, the field needs to be put into a highly specific context and tracked over time. That's obviously what university lecturers try to do. But even abbreviated versions can yield insight. When did the term originate? Why did specialist terms emerge at that exact point in time—what were the motivations of the early thinkers in the field? How is it defined and how has the definition evolved over time? Where does it fit with other academic subjects? What solutions does it provide? What are some known companies using these technologies and approaches? What are some findings by computer scientists recently? Most importantly, ask yourself: How can I use this domain to solve a

Figure 2.6 The field of computer science

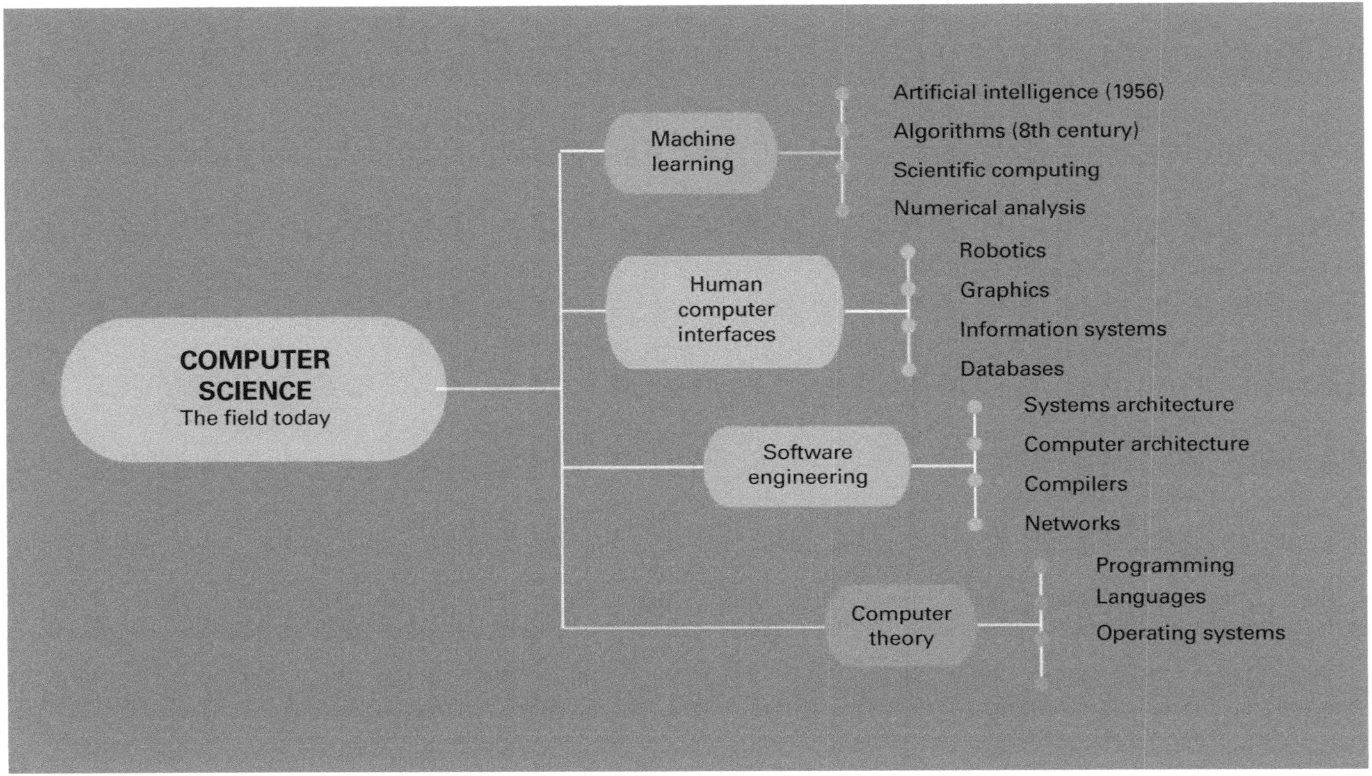

COMPUTER SCIENCE
The field today

Machine learning
- Artificial intelligence (1956)
- Algorithms (8th century)
- Scientific computing
- Numerical analysis

Human computer interfaces
- Robotics
- Graphics
- Information systems
- Databases

Software engineering
- Systems architecture
- Computer architecture
- Compilers
- Networks

Computer theory
- Programming
- Languages
- Operating systems

challenge I'm currently working on? Then, go ahead and solve that challenge—only by using it yourself do you fully understand a technology's potential.

Conclusion

Science and technology enable innovation in a myriad of ways, not the least understood yet. In fact, science and technology will each, in different ways, likely change more in the next decade than they did in the 30–50 years that preceded it. It is also true that some things that we may wish would change fast (e.g. the speed of creating suddenly needed vaccines or the path to understanding the depth of human emotions) might take much, much longer, either because they are exponentially more complex than the things we do make progress on, or because we haven't yet got the critical mass of pre-existing knowledge to build the required platforms of understanding to make true knowledge leaps possible or probable.

What we can observe is typically the results of *successful* innovations, which represent only a fraction of the activity going on at any given time. The challenge with that poor visibility of the whole activity is that it leaves us simply as consumers of that innovation, not as producers. Sci-tech innovations will affect all existing industries and will create new ones, which I will look at more closely in Chapter 4.

The only way to develop a sense of what's actually going on and have any idea about what might be coming down the pike is to attempt more creative approaches, such as the ones described in this chapter: discovering taxonomies, making tech trends your own, talking to scientists and technologists in person, being part of a community that cares about sci-tech (in some way), engaging deeply with (near any) topics in your surroundings topics on a daily basis in order to boil it down to what Elon Musk, echoing great thinkers such as Aristotle, Einstein, and Feynman, calls "first principles." First principles thinking, as opposed to thinking by analogy, or just memorizing known things, is the act of boiling a process down to the fundamental parts that you know to be true and building up from there, and, lastly, keeping a running written trail of what you are learning.

I shall now turn to a review of how policy and regulation moderate market conditions, and in so doing change the very foundation that science and technology had assumed was constant. Policies are not just reactive; they may set the stage for significant scientific development or, conversely, may inhibit some activities. Similarly, this can impact all other forces of disruption, including business and social forces, as I will study in further chapters.

Key takeaways and reflections

1 Reflect on your own relationships to science and technology. What are your strengths? What are your weaknesses?

2 What is your access to new science and technology? Do you have relationships to universities, experts, or startups? Do you have sufficient educational background (or raw intelligence) to be able to learn new things easily as they emerge? How would you do so? Online courses? Campus courses? Other methods? Searching the internet? Asking a friend? Does it depend on the topic? Then, challenge yourself: How could you become 3× or 10× or 100× better at it? Pick a tiny domain and do just that to prove to yourself that you can. Observe the difference.

   Try to run a quick analysis of four forces of disruption for a few science and technology topics that interest you at the moment, using Figure 2.7, the 4×4 matrix. If you cannot come up with any, try using artificial intelligence, Internet of Things, virtual reality, or blockchain. Identify at least three issues in each box. My advice would be to draw such a matrix for yourself at least on a monthly basis as a reflection exercise.

   Now, using Figure 2.8, try to assign disruptive emphasis to the topic you just considered. In other words, to what extent is this topic currently

Figure 2.7 Disruptive tech forces exercise

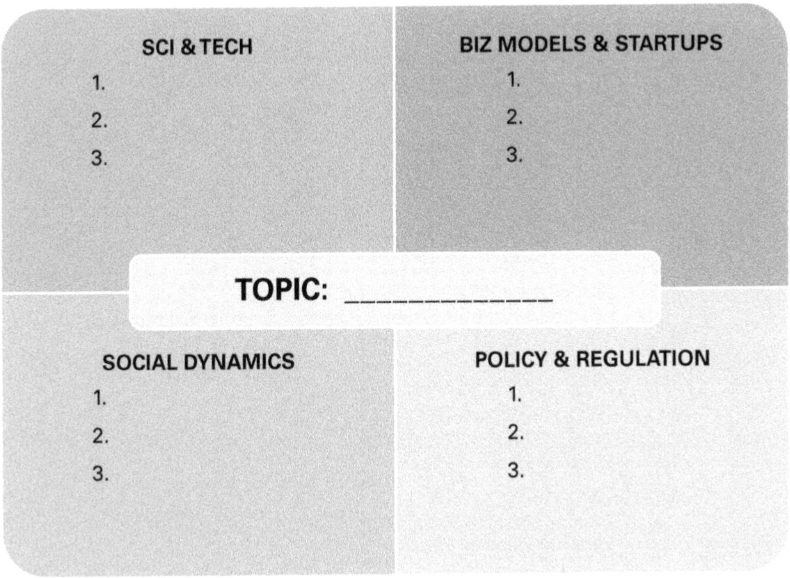

SCI & TECH	BIZ MODELS & STARTUPS
1.	1.
2.	2.
3.	3.

TOPIC: _____

SOCIAL DYNAMICS	POLICY & REGULATION
1.	1.
2.	2.
3.	3.

Figure 2.8 Disruptive emphasis exercise

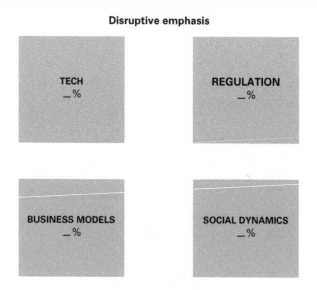

Figure 2.9 Visualizing technologies exercise

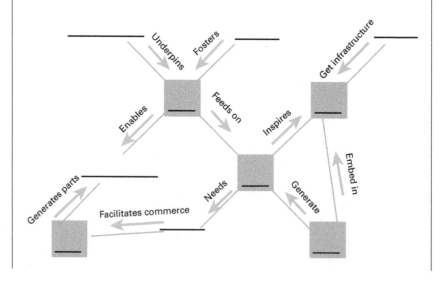

Figure 2.10 Technology overview exercise

SCI & TECH
1. Universities: _____
2. Research labs: _____
3. Scientists: _____
4. Other experts: _____

BIZ MODELS & STARTUPS
1. Biz models: _____
2. Startups: _____
3. Top events: _____
4. Industries: _____
5. Companies: _____
   Manufacturers: _____

TECHNOLOGY: _____

SOCIAL DYNAMICS
1. Consumption patterns: _____
2. Generations: _____
3. Other key trends: _____

POLICY & REGULATION
1. Emerging policy: _____
2. Existing legislation: _____
3. Adjacent regulation: _____

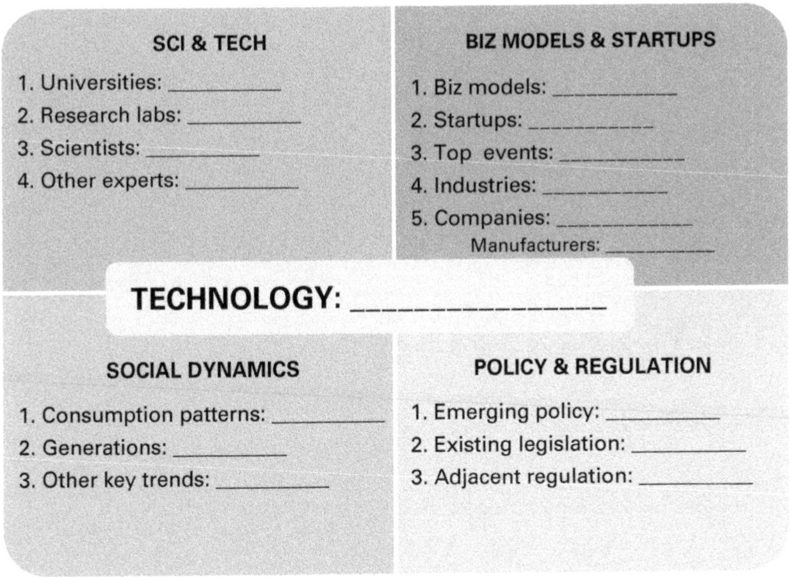

about the tech as such, or is it more about regulatory challenges? Or is the relevant issue which business model will emerge or prevail, or is it an issue of consumer adoption? Choose whether you will answer the question on behalf of yourself, your organization, or society at large. Draw arrows between topics that are related.

3   Try to think about how this will evolve, taking each year as an arbitrary unit (add more specifics if you can). Print out and fill a matrix for each year.

4   Fill in the blanks in the Figure 2.9 visualizing technologies exercise, using technologies that are relevant to your business, to yourself, or to society right now.

5   Fill in the blanks in the Figure 2.10 technology overview exercise and use it to compare with a colleague, if you can.

6   Develop some taxonomies of your own and compare them to some in this book or from other trustable sources. How do yours compare? What topics are unique?

Policy and regulation moderate market conditions

03

In this chapter, I look at the effect of government interaction in emerging markets and sectors that were created or are impacted by innovation in technology. The chapter has three parts: (1) I spell out the five roles of government (it questions, facilitates, scales, enables and, yes, restricts— but only in order to protect fairness); (2) I give examples of how the government plays one or more of these roles; and (3) I demonstrate government's specific role in technology innovation. Moreover, I discuss a few cases where the private sector takes on "government"-like roles, such as in voluntary standardization. The main message is that government is deeply involved with technology. As a large business using technology, as startup innovator, or an individual business professional using technology, you need to get smart about government. It will impact you in both positive and negative ways, so you might as well engage, attempt to foresee government action, or even influence it through the tools available to you. If you operate across markets, the exercise needs to be repeated per country and/or regional jurisdiction (e.g. the European Union).

The five roles of government

While many believe that government standardization, prohibitions, or de-tailed rules constrain and slow innovation, this is not all that government does. When operating at its best, by enabling a level playing field, govern-ment regulation also benefits equal market access, consumer protection, and public safety. I have rarely seen government functions defined this way, but the truth is, government has (at least) five challenging functions. The five roles of government (questions, facilitates, scales, enables, and restricts) each correspond to a specific policy instrument (conducting stakeholder consultation, writing policy documents setting out visions and plans for the future, implementing large-scale projects with high risk of failure, enabling innovation—or even innovating on its own, and creating regulations). Few organizations have such a complex mandate, which is why few outside of government have any (deep) understanding of why government quite often fails at executing all five equally well at the same time. No other type of organization would attempt to do so. There is an argument right there to simplify government's role. Yet, it carries out a fundamental function that is not so easily replaced by other actors (this is a longer discussion we are not entering into in this book).

Government forces can have different emphasis depending on the goal and the stages in a consultative process. In fact, you could look at this through five modes: consultation (questioning mode), policy (facilitating mode), innovation (enabling mode), risk taking (scaling), and regulation (re-stricting mode) (see Figure 3.1). For each of these, you can choose to study how each technology is handled individually, or you can look at it as a more complex matrix. Governmental culture also varies between countries, agen-cies, and administrations.

Figure 3.1 The five roles of government

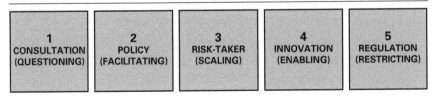

| 1 CONSULTATION (QUESTIONING) | 2 POLICY (FACILITATING) | 3 RISK-TAKER (SCALING) | 4 INNOVATION (ENABLING) | 5 REGULATION (RESTRICTING) |

1. Government as questioner

The government's first role, before making policy, is to consult its citizens and affected stakeholders—organizations, citizen groups, trade associations, and businesses—to figure out what to do next or how various ideas on next steps would affect the overall system they are elected to protect. This type of consultation, when done well, occurs throughout the policy process. Technology makes this more feasible and less expensive. Technology is also the reason why government has to act more often than before. Intervention is needed because society is changing.

The EU is a governmental entity that has perfected the role of public consultation. Every piece of legislation is out for consultation, and social and trade organizations comment actively. The end result is better policy because the stakeholders who would be affected are involved at the outset. This does not mean that they are always happy with the results.

Tracking country-based policy priorities is crucial for global businesses. Although there are websites that track some of that information, including the CIA World Factbook, many use consultants to get up-to-date advice in specific policy areas. Global firms also want to influence government tech policies. To do so, they conduct lobbying and work through industry associations. Major global industry associations in the technology field include CompTIA, BIO, and innovation attaché networks. Those organizations tend to be a good source of information about what hot policy topics are at any given moment, at least those that concern large firms. CompTIA, the world's leading tech association, is active on the tech topics of AI, blockchain, computer networking, cloud computing, Internet of Things, cybersecurity, drones, smart cities and virtual reality as well as the tech workforce. It is really an IT industry association. Board of director members at the time of writing include HP, Comcast, SAP, Cisco, Sophos, and EY.

Biotechnology Industry Organization (BIO, bio.org) is the world's largest trade association representing biotechnology companies, academic institutions, state biotechnology centers, and related organizations across the United States and in more than 30 other nations.

The American Association for the Advancement of Science (AAAS) advocates for behalf of federal funding for scientific research, aiming to be "a united voice for scientific progress." With a $100 million annual budget, AAAS is among Washington, D.C.'s best resourced and most influential advocacy groups (Nisbet, 2015). The organization publishes *Science* magazine. Current advocacy areas include STEM careers, diversity, human rights, ethics, science education, evidence-based policy making, and science policy, including the science budget.

Science Europe (scienceeurope.org) brings together 36 national research funding agencies and prominent research-performing organizations from 27 European countries. Among Europe's major players in public research funding, together they spend more than €18 billion on research each year.

For example, gene editing has been through different cycles through the years, going from a possibility which was supported by government R&D (enabling innovation), to actively facilitating it through policy intervention during the global Human Genome project through consultation (nationally and internationally). In the US, there are various state and federal bioethics commissions. The WHO Expert Advisory Committee on Developing Global Standards for Governance and Oversight of Human Genome Editing (the Committee) was set up by the Director General of the WHO in December 2018. Numerous policy issues arise with genome editing, insurance coverage and reimbursement of genetic tests, genetic discrimination, health disparities, human subjects research, informed consent, and intellectual property rights (to genetic data and DNA). The US Supreme Court (2013) has already decided that DNA as such, in its natural form, cannot be patented. That hasn't stopped companies from patenting a host of synthetic DNA variants, including the essence of a rose. Boston-based Ginkgo Bioworks has a synthetic bio-derived version of "rose oil" commercialized by leading essentials oil producer Robertet since 2017.

2. Government as facilitator and limiter

Nearly as important as questioning next steps is government's role as facilitator of discussions about where common priorities should lie, making and executing the resulting policies, and ultimately limiting the ability of various actors to circumvent the policy through regulation, thereby protecting both citizens and stakeholders. Its regulatory function acts a bit like a limiter in a digital audio system, constraining the end results to a manageable variability without excesses that distort.

Innovators in regulated industries such as biotech, transportation, and telecommunications are more used to considering the panoply of governmental forces than innovators in unregulated ones like the software industry.

However, as digitalization more significantly and deeply impacts all sectors, it forces some level of government scrutiny to return to previously regulated sectors as well as to the software sector itself.

At best, policy and regulation moderate market conditions in a positive way, correcting excesses that create inequality and market distortions. The biggest challenge in that regard is to ensure that policy makers are tech literate enough to make informed judgments. This is not an insignificant challenge, given the rapid evolution of new technologies, and the (arguably) historically slow learning curve of politicians and legislators when it comes to the same. When working for a parliamentary technology assessment organization, The Norwegian Board of Technology, it was sometimes striking how simple we needed to make our arguments to be heard. Many of the questions posed by policy makers were surprisingly naïve.

Furthermore, at the EU, I've been on both sides of the fence. As a policy maker and bureaucrat on e-government, I was faced with industry participation and was subject to meetings where they tried to "PowerPoint us" into a higher level of awareness of important issues in technology innovation.

As a lobbyist and strategist in big tech (Oracle) and working through trade associations and directly, I had the mixed blessing of undertaking significant efforts to sway EU bureaucrats toward our joint (and nuanced company-specific) industry view on technology standardization and interoperability. None of these activities was a completely innocent, pure teaching and learning endeavor. Neither were they "only" lobbying without concern for the "truth." Either way, my experience has exposed a dire need to dive deeply into the heart of the matters, which means exposure to all key debates, concepts, players, and issues over extended periods of time. Complexity is not going to go away. The age of the "common sense" politician who does not need to engage in the technical discussions is definitely over. With that kind of approach, one is doomed to make rash judgments, misunderstand key issues, and do more damage than good.

Taking a closer look at the elements of policy and regulation

The government has a plethora of instruments at its disposal. In this chapter we focus on policies, legislation, instruments, and stakeholder consultation programs to support technology. In Figure 3.2, I have added granularity to the elements of each.

Figure 3.2   The governmental forces of policy and regulation

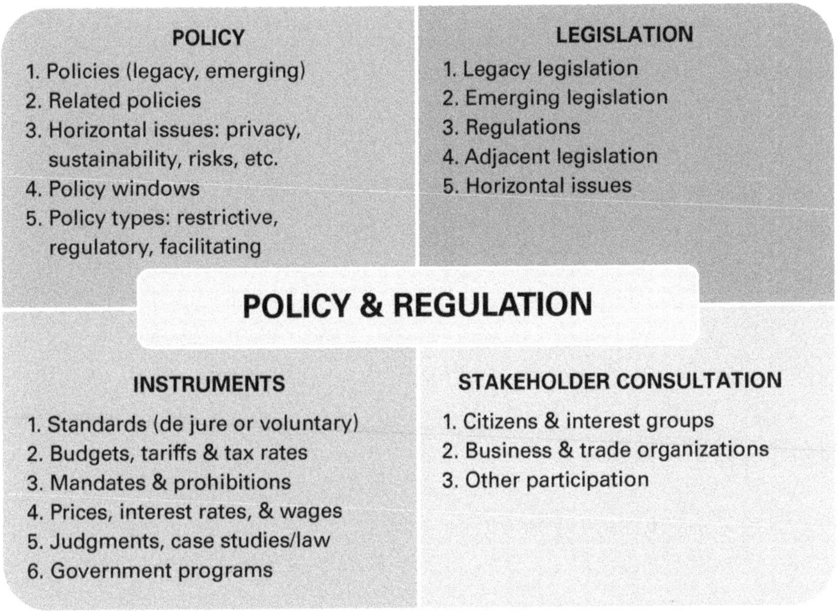

POLICY	LEGISLATION
1. Policies (legacy, emerging) 2. Related policies 3. Horizontal issues: privacy, sustainability, risks, etc. 4. Policy windows 5. Policy types: restrictive, regulatory, facilitating	1. Legacy legislation 2. Emerging legislation 3. Regulations 4. Adjacent legislation 5. Horizontal issues

POLICY & REGULATION

INSTRUMENTS	STAKEHOLDER CONSULTATION
1. Standards (de jure or voluntary) 2. Budgets, tariffs & tax rates 3. Mandates & prohibitions 4. Prices, interest rates, & wages 5. Judgments, case studies/law 6. Government programs	1. Citizens & interest groups 2. Business & trade organizations 3. Other participation

For example, tech policies (or regulation or legislation) cannot just be studied as finished products (they always will operate in a context of legacy policies, current policy, and emerging policies). They also cannot be viewed in isolation—rather, a multitude of issues will be tackled or joined up in any new policy proposal, potentially amending previous policies in horizontal or adjacent domains. Doing so requires collaborating across government—and consulting a wide array of actors. Similarly, when looking at the precise instruments in the government's arsenal, standards reign at the top of the list when it comes to technology. Next come budgets and tax rates. Government programs (R&D support, e-government initiatives, etc.) are often a source of initial momentum for a new technology (e.g. DARPA's support for the internet), or could be the push needed to make it part of national infrastructure (e.g. rolling out national electronic ID the way they did in Estonia). Throughout the chapter I touch on some of these elements, although going in depth on each of them would constitute a book in and of itself.

However, budgetary instruments go far beyond that. In 2021, the United States Federal government budget has allotted almost $53.36 billion toward the country's annual civilian federal agency information technology budget, not including military spending (Holst, 2020). There is also a host of mandates and prohibitions.

3. Government as risk taker (scaling)

The third role of government in the context of technology is in scaling efforts, which is inherently a risky endeavor. Government's engagements with technology at times fail spectacularly. Reasons are complex, but the sheer size of government tech infrastructure, its antiquated public procurement practices that lead to overpaying for highly simple tech that takes far too long to implement, and the struggle bureaucrats understandably have with keeping up with a field that is not their own, as well as all the lobbying which makes truth and facts hard to come by, all contribute to this reality. Also, I would like to see any organization implement a complex billion-dollar e-government program in health or social security, and in fact there are typically private contractors executing these programs, so part of the blame has to be put on them. The real reasons for the prevalence of failure and large cost overlays are complexity and a (somewhat misguided) implementation of bureaucracy to safeguard fairness, not incompetence as such.

One might argue that the US government's lack of COVID-19 tests in the first five months of the pandemic was one such surprising failure, given the track record of tech prowess of other parts of the US government. However, when more tech-savvy governments, such as Israel, China, Singapore, use surveillance tech to track COVID-19, this opens up a can of worms in terms of privacy (Kharpal, 2020).

As of late spring 2020, the US government got involved in an attempt to use smartphone data to track COVID-19. One prospect is that the CDC could use the info to track compliance with stay-at-home orders, or, perhaps more fruitfully, give early warning of potential new outbreaks based on mobility patterns (Lyons, 2020).

4. Government as tech innovator

The fourth role of government is that of a tech innovator, enabling progress to take place through supporting and financing innovators at the early stages of R&D. As an example of how government plays the role of fostering innovation, government funding was behind Google's search engine, GPS, supercomputers, AI, speech recognition, the internet, smartphones, shale gas, seismic imaging, LED, MRI, prosthetics, HIV/AIDS breakthroughs, and many, many more technologies (Singer, 2004).

However, the insights gained from successful engagement with technology tend to be stuck in small pockets of government and are not widespread—and get lost when the boundary-spanning innovators leave government for more lucrative jobs elsewhere. Government is also not great at proselytizing its own role. Indeed, sometimes they prefer it is now known. This is the case with intelligence or defense spending.

For example, most investments made by the US intelligence agencies' own venture capital firm, In-Q-Tel, are secret. Its mission is to identify and invest in companies developing cutting-edge technologies that serve United States national security interests. I have talked to them many times, but unless you co-invest or are being invested in (or considered for investment), you would never know what they are up to. Having said that, the website does include a list of (some) of their investments but does not specify that this represents their portfolio or what type of engagement these investments entail. Many companies listed on In-Q-Tel's (https://www.iqt.org/) investment website page are secret, and the products it has and how they are used is strictly secret.

When government plays the role of innovator it does several things at the same time: finances R&D, makes big purchases, and stimulates innovation through venture investment or by extending loans. This activity benefits both small and large companies.

As for exploiting tech for government (e-government), the northern part of the EU (the Nordics, the Netherlands), the UK, and a few select Asian countries and city-states (South Korea, Singapore, Hong Kong) have been ahead of the US according to most rankings (UNDESA, 2018). COVID-19 has added a few imperatives which are likely to be widely attempted (Gurria, 2020). First, improving connectivity, since broadband demand has soared and now underpins even basic business and education requirements, making the years following 2020 the time to overcome the digital divide, notably in rural and poorer households in particular parts of the country and poorer towns and villages.

Second, data access and sharing have become more important than ever, not just to share disease data but to overcome supply chain issues, collaboration, and coordination internally and across borders. What follows as the other side of the coin is a necessary focus on digital security, given that malicious actors are starting to exploit novel activity and novices online.

Third, improving key transactions that by now should be available online (certainly ecommerce, teleworking, and e-learning, but possibly also e-participation, online voting, and government service provision) is also going to be a priority. This challenge is definitely not just about equipment and technical capacity but also about improving skills. However, each of these has been part of ambitious government targets for decades, and they are not easily achieved and are prone to scandals and failure due to the size and complexity of the endeavor.

Regulating tech is not popular in the tech industry. Regulating business is not popular in the business community. Generally.

Regulation tends to be a political tool that appeals more to the left than to the right unless it is regulation to simplify government and reduce its size. As such, it will continue to be controversial and cannot solve all problems. When it swings too much in one direction it tends to be reversed in the next election cycle. Therefore, any attempt to regulate needs to take this dynamic into account. Having a long-term view on regulation and policy making is rare among politicians because it lacks incentives. Perhaps those incentives need to be adjusted to allow that kind of long-term view? Should more branches of government be elected or appointed for longer periods of time? Or should there be age quotas to ensure young people always are part of decision making (might one assume that a younger person takes a longer perspective since they are likely to live longer?). When we take into account the future evolution of tech and society, the future of government regulation is fascinating. What we cannot do is conclude that it will go in one particular direction. There is no data to help us out here.

5. Government as protector

The fifth role of government is to protect—not just its people alive and of voting age at any given second (voters and citizens) but the very foundation of society (its legacy and future). The protective role can be carried out through regulation that restricts certain activity or simply through monitoring that alerts people of activity that needs attention. Aside from that, government usually steps in when tech truly affects life and death issues. One case in point are the rules around the safety and procedures for clinical trials for new drugs. Some, both pharma giants, startups, and consumers awaiting new drugs, might want clinical trials to go faster and be less onerous, but nobody is arguing to get rid of that step. Besides, sometimes, in times of crisis, whether it is an unprecedented economic crisis or an external force such as a pandemic, or both in tandem, the circumstances call for regulation. When done correctly, the tide of regulation lifts all boats. When it goes wrong, regulation sinks the wrong boats and creates stormy waters for everyone. Either way, government regulation of technology can only be ignored at your peril.

Europe is a global trendsetter on tech regulation, perhaps due to the nature of its construction, a constellation of 27 member states with separate markets gradually being brought into one while keeping languages and local flair. A study found near half of regulations proposed in 2019 originated from the region (Murgia, 2019). For example, the automotive industry is globalized through the Global Technical Regulations, which were initiated by the European initiative the World Forum for Harmonization of Vehicle Regulations (1952) and currently are developed under the 1998 international agreement on vehicle construction. The regulation concerns lighting, controls, crashworthiness, environment protection, and theft protection. Yet, cars in the EU, the US and Asia also have regional requirements, and the US and Canada were not part of the 1952 or 1958 agreements, although they did take part in the 1998 agreement. Global tech regulation is complex because huge, nationally flavored business interests intervene.

Censorship

Governments do tend to get involved with censorship at various stages. Typically, this is the story of China, but many more countries practice internet censorship from time to time. In the US, freedom of speech is protected by the First Amendment of the Constitution, with the primary exceptions being obscenity, including child pornography, defamatory speech, internet content that poses a threat to national security, the promotion of illegal activities such as gambling, prostitution, theft of intellectual property, hate speech, and inciting violence, all of which do not enjoy First Amendment protection. After 1998, the Digital Millennium Copyright Act (DMCA) criminalizes the production and dissemination of technology that could be used to circumvent copyright protection mechanisms and limits the liability of the online service providers for copyright infringement by their users. The latter is now suddenly up for discussion again. If the proposed EARN IT Act were passed, tech companies could be held liable if their users posted illegal content (Newton, 2020).

Internet blocking tools tend to get used by governments to solve public policy issues, using three main methods: IP and protocol-based blocking, URL-based blocking, and deep packet inspection-based blocking, yet they are scarcely effective (Internet Society, 2017). Having said that, a renewed focus on paywalls and gated access among both commercial providers and governments in order to use the internet as a communication tool for non-public content is rendering the extreme position of no blocking equally untenable.

Around the world, Facebook censors content in order not to be banned by various governments, including Germany, India (if users criticize religions or the government), Russia (content about Kremlin critics), Turkey (if users depict Muslim religious figures or poke fun at the nation's founder), and Pakistan (if users claim a cozy relationship between the government and Islamic militants) (Paglieri, 2005).

Government's role in regulating technology

Three examples where government plays a particularly active role are in antitrust, drug development, and standardization. I have the following examples in mind: Congress/antitrust on big tech, 23andMe, and HIPAA standardization.

US Congress and regulation of big tech

A similar debate is now occurring with regards to regulating big tech (Facebook, Apple, Alphabet, Verizon), after lawmakers in Europe and the US are set to unleash an array of new legislation to curb the growth and influence of big tech on government, business, and society (Markman, 2019). FAAMG—Facebook, Amazon, Apple, Microsoft, and Google (Alphabet)—have a combined market capitalization of more than $4 trillion (Jones, 2019).

There are several ways antitrust efforts can play out. Breaking up big tech is not the only option and not necessarily the approach with the best result, as simple as it seems. In the EU, too strict enforcement of competition law will negatively impact the development of national champions, which is a true and tested approach which most member states have been pursuing for years. US antitrust law makes an aggressive approach by government very tough, given the burden of proof requirements on the plaintiff (Aridi and Urška Petrovčič, 2020). The goals of antitrust are also different. In the US, the goal is consumer welfare, but in the EU there is also a strong focus on market integration and removing artificial barriers to trade. In either economy, any attempt to change the status quo faces strong lobbying efforts on either side of the argument. Notably, big tech players that somehow escape scrutiny this time around are likely to try to bolster support for antitrust against their competitors.

The aftermath of the infamous General Data Protection Regulation (GDPR), which was adopted by the EU in May 2018, included an increased emphasis on data privacy but also major hurdles for small and large businesses which are involved in digital advertising, and it has complicated even basic things like sending out newsletters to subscribers, putting the burden of proof on the firms.

The case for regulating big tech is that monopolistic markets are less innovative and tend to box in consumers, giving us less choice and higher prices. The case against regulating big tech is that such companies are capable of releasing a stream of innovations at global scale, which provides massive opportunities for their supply chains and generates consumer products that the majority enjoy. Breaking them up or changing their ability to market certain products interferes with their commercial operation and might slow them down or, at worst, provide less of an incentive to innovate. I'm not aware that there is a generic answer to which is the truest description of where we are at this historical moment. Arguably, some markets that have been dominated by a single player, such as search engines, are overripe for intervention. On the other hand, that market is now showing signs of self-correction.

US government regulating medical disclosures and claims

The startup 23andMe provided an inexpensive test that would reveal genetic predispositions for dozens of traits. In 2013, Anne Wojcicki, founder of 23andMe, didn't think she needed approval to provide information about her customers' health risks around genetically affected diseases including cancer. The US Food and Drug Administration (FDA), predictably, had a different view (Hayden, 2017).

What followed was years of battling with the agency before the company was allowed to operate again, although not with free rein. After two years, it applied to disclose the results of one test, for a rare disease, which was approved. More approvals followed. It still cannot disclose all its results for allegedly still controversial tests such as those connected to the BRCA genes, which are said to be linked to breast cancer.

Whether the brazen approach taken by 23andMe to go ahead without approval was deliberately naive or not, it did cause significant challenges and almost broke the company. Startup innovators would be wise to solicit counsel way before this kind of move, especially in a highly regulated industry.

There are important reasons why consumers should be protected from a random company's health pronouncements about your genes. The domain is, in fact, arguably the epitome of necessary and productive government regulation. Yet, it seems clear that if government takes years to approve the release of important new information pertaining to genetic disease, that is equally problematic on a societal level. As with many things in life, it is a game of checks and balances.

The precautionary principle is often invoked when it comes to life science innovation. It's a practical as much as an ethical consideration, particularly in Europe. It is politically contentious, especially in the US, where a broad notion of individual freedom stands very strong.

Government influence on cybersecurity (HIPAA)

Let us look at a different domain: that of the workplace, where safety and health are closely tracked by governments worldwide. Convincing workers to work remotely is not the only challenge for employers. There are significant legal issues with remote teams. Privacy and security must be ensured both in terms of the customer data shared and the data on employees. Cybersecurity is a growing concern. Payroll must take account of the state in which the worker is registered. Reasonable accommodations must be made so that the new remote workplace does not discriminate against any workers.

For many companies and government functions, cybersecurity is a major obstacle to remote work. At home, WiFi networks are typically not as secure as office networks, and personal computer equipment (desktops, laptops, printers) are also typically less secure and can accessed by others. As video-conferencing startup Zoom became the default remote work solution for business and education almost overnight in spring 2020, stories of hackers appearing on its platform also appeared. The platform was simply not prepared for the increased attention and scrambled to respond. However, Zoom does now offer a Telehealth plan for business accounts which is HIPAA compliant, where meetings are encrypted end-to-end, and Cisco Webex has done so since 2018. Cloud recordings are disabled but are enabling health personnel to provide remote patient care.

Existing government regulations for privacy and cybersecurity in the US include the Health Insurance Portability and Accountability Act and its Privacy and Security Rules (HIPAA), the Gramm-Leach-Bliley Act (GLBA), which governs non-public personal information, the Office of Compliance Inspections and Examinations of the Securities and Exchange Commission (SEC and OCIE), which governs SEC-registered entities including broker-dealers, and the National Institute of Standards and Technology (NIST) controls and standards (Orrick, 2020).

Tempting as it is, many underestimate governments. Contrary to what one might think, there is an intimate relationship between government and technology. The DARPA program financed the internet. Government space programs such as NASA dominated the space race for 50 years until the private sector took up the mantel. The EU framework programs for research have financed breakthroughs in R&D (and less commercial solutions). Government procurement is the lifeblood of many industries, including defense, transportation, telecommunications, energy, software, and manufacturing.

Government standardization of technology

Governments have a particularly strong role when it comes to standardization. There is hardly any technology where there is no governmental standards body controlling its development or at least watching over it. The aim is to create a level playing field for competition and ensure equal access for government itself, and for businesses and citizens. You could say this is the paradigm case of the government's protector role on behalf of the public.

The International Organization for Standardization (ISO) was founded in 1947. It is headquartered in Geneva, Switzerland, and has 164 national standards organizations as members. Each member represents its country's standardization activities to ISO and, in turn, represents ISO back to its own country. ANSI represents the United States. There are currently 23,217 ISO standards and about 10,000 American National Standards, but it is a semi-official body in the United States.

ISO defines a standard as "a document that provides requirements, specifications, guidelines or characteristics that can be used consistently to ensure that materials, products, processes and services are fit for their purpose." ISO's marketers have found a quick-hand description and now encourages us to "think of them as a formula that describes the best way of doing something" (https://www.iso.org/standards.html). The most well-known standards include the ISO 9000 family of quality management standards, the ISO 27001 information security standard, and ISO 3166 country codes, which allow us to painlessly refer to countries in a uniform way.

The ISO 9000 standards family is based on a number of quality management principles, including a strong customer focus, the motivation and implication of top management, the process approach, and continual improvement. There are more than 1 million companies and organizations in over 170 countries certified to ISO 9001. Using ISO 9001 helps ensure that customers get consistent, good-quality products and services (ISO, 2020).

ISO/IEC 27001 is widely known, providing requirements for an information security management system (ISMS), though there are more than a dozen standards in the ISO/IEC 27000 family. Using them enables organizations of any kind to manage the security of assets such as financial information, intellectual property, employee details, or information entrusted by third parties.

An example of how the ISO 3166 country code standards is applied to technology is how the internet domain name systems use the codes to define top-level domain names such as ".fr" for France and ".au" for Australia. If you think about it, when standards work and are implemented, they seem incredibly simple. It is when they do not work that the problems start. Imagine the confusion if countries were regularly confused with each other when sending mail or email. Consider the countries Slovakia and Slovenia. If you did not have a clear standard, who would know which was abbreviated SL and which was abbreviated SK?

The business benefits of standards

For business, standards are strategic tools that, when successfully deployed, have a number of benefits to themselves and to the clients they serve. Standards can reduce costs by minimizing waste and errors. Standards can increase productivity by streamlining processes, making them more efficient. Standards can enable network effects from providing common approaches that have the potential to enable technology platforms

Having said that, being an early proponent or adopter of a standard rarely gives lasting competitive advantage, unless it is a proprietary standard (also called a *de facto* standard), which is controlled by one company alone. Examples of this would include the Word document format that formed the core of the Microsoft monopoly on desktop productivity tools in the 1990s, securing the dominance of the Office suite. Another example is the iOS and Android ecosystems, which I discussed at some length in the previous chapter.

A standards-setting organization (SSO) is an organization charged with developing, coordinating, promulgating, revising, amending, reissuing, interpreting, or producing technical standards intended to address the needs of a group of affected adopters.

Most standards are voluntary in the sense that they are offered for adoption by people or industry without being mandated in law. Some standards become mandatory when they are adopted by regulators as legal requirements in particular domains.

Private, voluntary standards shape almost everything we use, from screw threads to shipping containers to e-readers. They have been critical to every major change in the world economy for more than a century, including the rise of global manufacturing and the ubiquity of the internet.

Voluntary standards setting

Up until this point, I have discussed government-organized standards setting. However, voluntary standard setting in an organized form dates back to the 1880s in Europe and the United States. Committees of engineers that were initially typically sponsored by engineering societies and later by standards-setting organizations began creating standards that would be adopted by manufacturers to satisfy the needs of corporate customers.

As Yates and Murphy (2019) point out in *Engineering Rules*, standardization is a process with an astonishingly pervasive, if rarely noticed, impact on all of our lives. The paradox is that the private sector voluntarily performs self-regulation, e.g. it takes on the role governments would otherwise do. The benefit is, of course, that it then has greater control over what regulation passes.

"By the 1920s, standardizers began to think of themselves as pivotal to world peace and global prosperity," writes Yates. "Container standardization helped make possible today's global supply chains ... Some economists say that container standardization has had a bigger impact on the enormous growth in global trade than anything the World Trade Organization has done" (MIT Sloan, 2019). Intermodal shipping containers—those big metal boxes that transport goods on ships, railroad cars, and trucks—are the crucial element in an efficient global supply chain because of Swedish engineer Olle Sturén, the long-serving ISO Secretary General who was a standardization pioneer. He led the Swedish delegation to ISO which created the 40-foot container standard in 1968 (ISO 668), following the American truck size and forcing various European truck and rail transport systems to change. Ultimately, the global economies of scale from this simple decision are almost incalculable and have led to the current explosion in global freight and supply chains.

Voluntary standards setting continued even after the establishment of ISO, particularly in the information technology field, where privately funded standards organizations called "consortia" emerged around 1987 as a result of a perception that government standardization was too slow and cumbersome, and they continue to exert considerable influence over the global marketplace. There are more than 500 technology consortia in the world today (https://www.consortiuminfo.org/). Most of them are run by members representing companies, although government and universities might at times join, but individual end users usually cannot. There is an obvious need for products to plug and play together. In fields where products emerge rapidly, such consortia can appear literally at a moment's notice.

Standardization of systems of innovative technologies (GSM, UMTS, WiFi, DVD, BlueRay, MPEG, etc.) involves a mix of rivalry, coordination, and collaboration. In fact, consortia are typically formed in situations where the ownership of IPR is fragmented among several market participants and when the level of technology rivalry among firms is high (Baron and Pohlmann, 2013), but even though consortia might improve the efficiency of R&D by coordination, they may simultaneously reduce competition. Three of the top standards consortia for information technology include the IETF, the W3C, and OASIS. I'll briefly explain what they do.

The Internet Engineering Task Force (IETF) formulates, publishes, and regulates internet standards, particularly those related to TCP/IP. The World Wide Web Consortium (W3C) is a community of a large number of member organizations, which work together to develop web standards and improve web services. Popular standards developed by W3C are HTML, HTTP, XML, and CSS, each of which is actively used by millions every day.

The Organization for the Advancement of Structured Information Standards (OASIS) is a global nonprofit standards body (e.g. "consortium") that works on the development, convergence, advisory, and adoption of open standards for cybersecurity, blockchain, privacy, cryptography, cloud computing, IoT, urban mobility, emergency management, content technologies, and more in order to foster interoperability. Some of the most widely adopted OASIS standards include AMQP, CAP, CMIS, DITA, DocBook, KMIP, MQTT, OpenC2, OpenDocument, PKCS, SAML, STIX, TAXII, TOSCA, UBL, and XLIFF, many of which are crucial to the operation of business or government transactions. OASIS can trace its origins back to 1993. Given its track record working with ISO, it advertises itself as a path to de jure approval for reference in international policy and procurement.

Standards promote interoperability which opens up the market

The main reason standards work is that, when correctly developed, they enhance interoperability. Interoperability is the ability of different information systems, devices, and applications to work together seamlessly within and across organizations and products. Without interoperability, you risk getting locked into specific solutions, which narrows your choice and may force you to keep purchasing solutions from a single vendor in perpetuity. Conversely, platform companies typically work hard to achieve this very situation. For that reason, the quest of interoperability is a constant battle.

Standards enable innovation

Without question, successful standards that are implemented by most vendors enable innovation. A good example is TCP/IP, which is used to govern the connection of computer systems to the internet. Without such a standard, there is no internet. Standards boost transparency—the example is PDF, which enables the sharing of documents. Standards can help avoid lock-in—the example is ODF, which is an open document standard that now has replaced proprietary Microsoft Word formats for the exchange of editable documents in many governments around the world, which enables many providers to provide document editing software products, allowing a free marketplace. Standards can help create market stability—the example is HTTP, the basic principle behind communication for the web, where hypertext documents include hyperlinks to other resources that the user can access. Standards can ensure efficiency and economic growth—the example is the internet, which arguably is the backbone of economic growth today.

While regular fora-/consortia-based standards activities are club goods (in economists' lingo) since they are not fully available for the common good, some players are not content with that and instead form cartels. The most recent high-profile example might be the clean emission technology collusion among automakers set up to fool EU regulators. The case opened in 2018 and includes BMW, Daimler, and VW, the major German automakers (EC, 2020). Sports media rights is another area of concern.

The key policy areas to watch out for

Technology policy is no longer a sector policy that can be buried in a debate away from mainstream policy developments. By 2020, it belonged squarely among the top three of fiscal, trade, and technology policy, with sector policies as a distant fourth, depending on the strategic importance of the sector in each jurisdiction, of course.

There are five overlapping policy domains that affect how governments regulate technology: fiscal policy, trade policy, technology policy, industrial policy, innovation policy, as well as a set of constantly overlapping sector policies (such as separate industry policies, health policy, worker regulation, and more).

Governments' science and technology (S&T) budgets tend to shrink or increase along with gross domestic product growth in similar pace with total government expenditure (Makkonen, 2013).

Government's role in social dynamics—the guardian of ethics

Finally, as the emphasis on technology increases, so should the focus on the ethics and global governance of emerging technologies. If not, we might inadvertently put in place a change that is touted as an improvement but where the full impact has not been sufficiently studied beforehand. This is also a continuous tracking challenge, since it's impossible to predict all negative fallout and we might have to mitigate negative consequences after the fact. See https://www.oecd.org/sti/science-technology-innovation-outlook/

How to track government regulatory activity across the world

To track emerging developments, a few institutions stand out as good sources of information, namely the OECD, EPTA (https://eptanetwork.org/), the parliamentary technology assessments organizations across Europe, and the innovation attaché networks that stimulate international cooperation between companies, knowledge institutes, and governments in the fields of innovation, technology, and science (Swissnex, Invest in Holland, Innovation Canada, Innovation Norway, etc.).

The OECD runs fairly deep tracking of innovation across the world. The in-depth OECD Reviews of Innovation Policy are particularly useful. Recent studies were made of Portugal (2019), Austria (2018), Kazakhstan (2017), Norway (2017), Finland (2017), and Costa Rica (2017). Each review identifies good practices from which other countries can learn.

Parliamentary technology assessment evolved in the United States in the 1970s and throughout the 1980s and 1990s in Europe. It meant greater emphasis on the broader socioeconomic impact of technology (Joss and Belucci, 2002). Despite keeping a relatively low profile, EPTA's members pack a punch with the depth and breadth of their tech policy focus. Being focused on giving advice to parliaments, their reports are written in layman language and are always current with the political debate. When I worked for the Norwegian Board of Technology, we contributed to changing interoperability policy for document formats and shaped renewable energy policy. New projects among EPTA's members in 2020 included bioelectronics, face recognition, social distancing, digital mental health, gene editing, de-confinement, and biometrics (EPTA, 2020).

The ethics of emerging tech

Given the societal consequences, there will have to be tracking and timely policy interventions on all important emerging technologies, such as quantum tech, AI, nanotech, synthetic biology, and each of the technologies outlined in Chapter 2. These technologies will, in turn, offer new opportunities as powerful regulatory instruments (Brownsword and Yeung, 2008). One way we can envisage working at huge scale, in the future, would be tracking and monitoring of sites such as meat-packing plants. Previously, this required a physical inspection. Now, with the coronavirus threat being an additional concern for the inspectors, one would have to expect that online monitoring could become the norm, with the potential efficiencies that would bring (as well as ethical issues, of course). Tech tools available to do so would be plain old webcams, AI for indexing the image footage and tracking anomalies, as well as unmanned aerial vehicles (UAVs) for monitoring the outside environment (Eggers et al., 2018). If you add environmental sensors in an Internet of Things scenario you could reproduce the factory's control room remotely. Much of regulation is also parse-able as code and could be understood by machines directly, which could implement remedies on the spot. Until such technologies are fully understood or controlled, however, these uses might end up offering undiscriminating views of what's going on, leading to the kind of situation we have in 2020 around facial recognition, which misrepresents a huge percentage of black people, leading companies such as IBM to halt work on facial recognition (whatever that means).

Legislative tracking software is slowly improving and will enable easier tracking of ongoing and emerging initiatives, which is important given the sheer mass of activity. Best practice exchange also has its place. Back in the mid-2000s, I built out the EU's biggest e-government best practice sharing framework, called ePractice.eu and now folded into Joinup.eu, taking it from 150 passive users to 150,000 active users across Europe. Sustaining such efforts over time, however, is challenging, as cases rapidly become stale and outdated as the sociopolitical context changes.

Country-based indices like the annual Bloomberg Innovation Index, WIPO's Global Innovation Index (GII), Transparency International's Corruptions Perceptions Index (transparency.org/en/) are quite helpful to track tech policy, although each index typically has its own problems and measurement biases. These indices tend to track general features of the regulatory environment (intellectual property rights protection, the independence of the judiciary, and the efficiency of the law-making process) as well as more ICT-specific dimensions (the passing of laws relating to ICT or the software piracy rates). Each of these indices is highly correlated but they are never 1:1 due to each using different metrics. The biggest difference is that some indices give points for country size (inadvertently or overtly) and others don't, which affects the position of smaller nations such as the Scandinavian countries as well as Singapore.

To show that regulatory initiatives are heavily dependent on the societal context, let me use the example of face recognition. As one of the areas where AI actually has made some headway, it is perhaps somewhat surprising that it has come under such scrutiny. Initially, COVID-19 should perhaps have worked in favor of more face recognition to fight the virus, but then came the demonstrations against police violence following the various murders of Black Americans in spring 2020. This cast light on the fact that face recognition currently fails in a racially disparaging way, disadvantaging Blacks. Until that is fixed, the technology could be considered as a racially biased technology, which suddenly and overnight makes it extremely contentious. However, facial recognition is only one among the many instruments that constitute the surveillance society that technology enables. Focusing on one aspect only will not solve that problem.

Ironically big tech is now calling for legislation of hate speech, disinformation, algorithmic discrimination, political ads, and more, largely because enforcing unclear rules and self-enforcement is not creating a level playing field (Sherman, 2020).

Emerging areas of tech policy

A few areas stand out as no-brainers for government intervention in the decade ahead: antitrust, privacy, consumer protection, censorship, digital services tax, supply chain security, and sustainability. As for specific emerging tech topics, look for regulation on all the emerging technologies mentioned in Chapter 2, specifically perhaps AI, healthcare IT, remote workforce, cybersecurity, Internet of Things, drug pricing, facial recognition, spectrum sharing, synthetic biology, renewable technology, human augmentation, germline gene editing, autonomous vehicles, blockchain, neurotechnology, additive manufacturing, quantum computing, and nanotechnology (Bernard, 2020; Kutler and Serbee, 2019; Marr, 2020).

The challenge with any attempt at regulating emerging technologies is twofold: one, they are global in nature and rely on platforms and networks that don't neatly fit into the nation-state regulatory frameworks of each country, and two, they emerge rapidly, which is antithetical to the stability and slowness of a bureaucratic regulatory process. That tension will not go away, even with international collaboration or even with a more efficient government regulatory process.

The deregulation movement sweeping across world governments in the last 25 years has created a sense that any new regulation presents an unnecessary administrative burden. The Dutch have perfected the analysis of administrative burdens and have moved quite effectively not just to measure such burdens but to gradually remove legislation with unacceptable burden. The Dutch government's 2003–2007 Administrative Burdens Reduction Programme generally gets high marks for aiming (and achieving) the quest to eliminate €4 billion of administrative burdens on business by 2007

(World Bank, 2007). The challenge with such an approach is where it should end. Also, what is left then when the pendulum inevitably swings back? If you undo too much legislation, bringing it back might bring back monsters you thought you were rid of. Gradual approaches don't carry the same risks.

Then there's the principle of caution or constraint, which says that any moral community (government or otherwise) needs to be careful that emerging technology does not undercut the very conditions of the society that community wants to foster (Brownsword and Yeung, 2008). To some extent, governments try to safeguard this principle by stakeholder consultation. In the EU, for example, stakeholder consultation is a formal process by which the European Commission collects information and views from concerned parties about its policies. The aim is to ensure participation, openness, effectiveness, and coherence (EU, 2020b). There is a considerable web of organizations represented in Brussels whose sole purpose is to respond to these consultations, both when the EU prepares and when it evaluates the impact.

Will self-regulation solve all problems?

The answer is in some cases to allow for self-regulation (for instance through aforementioned standards consortia or even less formally through trade association discussions and charters), but that cannot always be the answer as it is prone to being exploited. As stakes increase, a "neutral" arbiter needs to intervene. Except, no government or multinational institution is neutral in this sense. Each has agendas, whether it be protecting its citizens or defending more ephemeral human values across nations and attempting to codify those within some conception of a global rights paradigm, human rights or otherwise. In either case, both efficiency and legitimacy of the regulatory process are inherently complex, but so is its absence.

Conclusion

Policy and regulation moderate markets quite powerfully and the effect can be market enhancing, correcting, or downright detrimental to innovation. As many other things in our society, which effect it has depends on many factors. In this chapter, I pointed out how consumer DNA company 23andMe initially misread federal regulation and suffered a two-year backlash, and how remote work might enjoy an uptick because of coronavirus restrictions but it partly depends whether governments facilitate regulatory ease for businesses that want to do this more permanently. I then brought up the importance of standardization and ended with how and where exactly I envision big tech will be regulated.

Conceptually, I've described the way government forces tend to emphasize five action modes: (1) *questioning* (through consultation), (2) *facilitating* (through policy), (3) *risking* (through necessarily large-scale project implementations), (4) *enabling* (through R&D and innovation), and (5) *restricting* (through regulation). If we are more aware of when each mode is open for business or citizen input, we can influence the direction of the technologies we care about. To do so requires investment of time and effort, and learning about how to do so specifically, and may require organizing and mobilizing support through our various networks.

As professionals, we need to fully understand the impact of the technologies we produce or use. Once we are sure, we should feel free to take strong, bold positions to empower society with our solutions or criticisms. This chapter has given you some additional tools to conduct a critical assessment of where government regulation is going at any given time, and perhaps also ideas on how to intervene to empower government officials and technology users alike with a more complete understanding of what is going on. That awareness will only become more important as technology continues to permeate the fabric of our society.

Public discussion about tech governance is often hampered by lack of awareness of both the opportunities and the limitations of government intervention given tech complexity and change as well as the influx of new business models that regularly alter the playing field. Clearly, government officials at times lack sufficient overview of technology and certainly lack the foresight to predict where they should legislate until it is very late in the game. Similarly, many professionals have an insufficient understanding of how governmental forces work. However, at times, smart regulators and smart tech business professionals are outsmarted by politics of the moment, which may override any good intentions based on sound evidence. That's the price we have to pay for democracy and political leadership, which at times overwhelms semi-independent regulatory agencies across the world.

I hope to have corrected that by pointing to the clear concepts that guide such policies and by keeping the dignity of choice sacrosanct—since quality of life is built through invention and reflection, not through regulation (Brownsword and Yeung, 2008: 48). In the next chapter, I will pick up that discussion from the perspective of how new business models upend markets, whether enabled by new tech platforms or simply by a good idea.

Key takeaways and reflections

1 Reflect on your own relationship to policy and regulation. What is your first thought about how government affects innovation and technology?

2 What is your access to new policy and regulation? Do you have relationships to government decision makers, policy makers, or bureaucrats?

Fill in the blanks in the Figure 3.3 government forces exercise, using the regulatory situation for a technology that is relevant to your business, to yourself, or to society right now. Find specifics for each subcategory. If you cannot come up with any, try using artificial intelligence, vaccines or robotics. Identify at least three issues in each box.

Figure 3.3 Government forces exercise

POLICY	LEGISLATION
1. Policies (legacy, emerging): _____	1. Legacy legislation: _____
2. Related policies: _____	2. Emerging legislation: _____
3. Horizonatal issues: privacy, sustainability, risks, etc.: _____	3. Regulations: _____
4. Policy windows: _____	4. Adjacent legislation: _____
5. Policy types: restrictive, regulatory, facilitating: _____	5. Horizontal issues: _____

POLICY & REGULATION: _____

INSTRUMENTS	STAKEHOLDER CONSULATATION
1. Standards (de jure or voluntary): ___	1. Citizens & interest groups: _____
2. Budgets, tariffs & tax rates: _____	2. Business & trade organizations: ___
3. Mandates & prohibitions: _____	3. Other & participation: _____
4. Prices, interest rates, & wages: ____	
5. Judgments, case studies/law: _____	
6. Government programs: _____	

Assign percentages to Figure 3.4. Attempt a comparative analysis either over time (historically or future-oriented) or across countries for a technology you care deeply about and are engaged in. To what extent can you or your company influence current priorities? With what strategies and tactics?

Figure 3.4 Government forces emphasis exercise

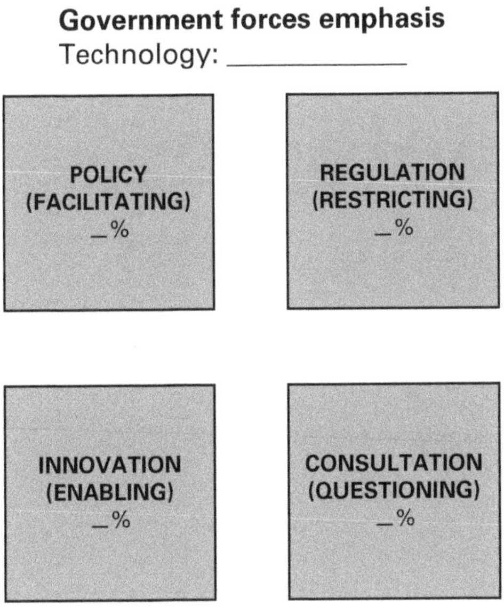

Government forces emphasis

Technology: _____

```
┌─────────────────────┐   ┌─────────────────────┐
│                     │   │                     │
│     POLICY          │   │    REGULATION       │
│  (FACILITATING)     │   │   (RESTRICTING)     │
│      _ %            │   │       _ %           │
│                     │   │                     │
└─────────────────────┘   └─────────────────────┘

┌─────────────────────┐   ┌─────────────────────┐
│                     │   │                     │
│    INNOVATION       │   │   CONSULTATION      │
│    (ENABLING)       │   │   (QUESTIONING)     │
│      _ %            │   │       _ %           │
│                     │   │                     │
└─────────────────────┘   └─────────────────────┘
```

3 Standardization: What are the most important standards impacting the business you work in? Which standards are emergent or need to emerge to evolve your sector?

How business forces upend technologies, markets, and society

In this chapter, I reveal how business forces have the capacity to upend not just markets but also technologies and, ultimately, society. Today, business models, among the most important disruptive subforces in the business domain, have become a major instrument of innovation (see Figure 4.1). The challenge is that the novel business model of today may not help us understand what will work tomorrow. Rather, we need to look for such novelty ourselves, to protect our business, to capture opportunity, and to understand how society itself is changing. In other words, we need to move well beyond business model analysis. I assist you in doing so by (1) providing the key context of the historical evolution of industry taxonomies (so you can be aware of specific socioeconomic patterns), (2) explaining how things have worked in the (immediate) past for technology-related businesses, and (3) looking at how overarching strategy frameworks can clarify choices and pathways as you aim to look 3–5 years and perhaps a decade ahead.

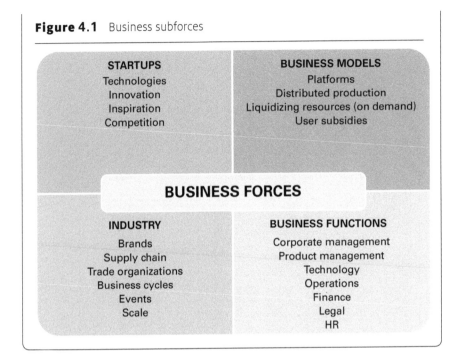

Figure 4.1 Business subforces

STARTUPS	BUSINESS MODELS
Technologies	Platforms
Innovation	Distributed production
Inspiration	Liquidizing resources (on demand)
Competition	User subsidies

BUSINESS FORCES

INDUSTRY	BUSINESS FUNCTIONS
Brands	Corporate management
Supply chain	Product management
Trade organizations	Technology
Business cycles	Operations
Events	Finance
Scale	Legal
	HR

How industry taxonomies illustrate technology change

The idea that there are different industries with separate concerns (and un-equal dynamics) is a long-standing notion in our society, dating back to the industrial evolution from which many of those industries gradually emerged. Schumpeter, the leading theorist on industrial change, claims that as the creation of new industrial activities is made possible by technological pro-gress, creative destruction, enabled by innovation efforts, is a key part of the process leading to the emergence of new sectors (Schumpeter, 1942). These disruptions are fueled by entrepreneurs. However, the process is cyclical.

As entrepreneurs become successful, they enter the capitalist class and their creations become corporations and as a result powerful and capable of im-mense scale—with Google and Amazon being recent cases in point—vulnerable to the same weaknesses, including group think, complacency, and slowness to change, as other corporations before them. As Jeff Bezos alluded to in a famous letter to staff as recently as 2018, creative destruction eventu-ally (and without notice) catches up with everyone, and "I predict one day

Figure 4.2 Industry timeline

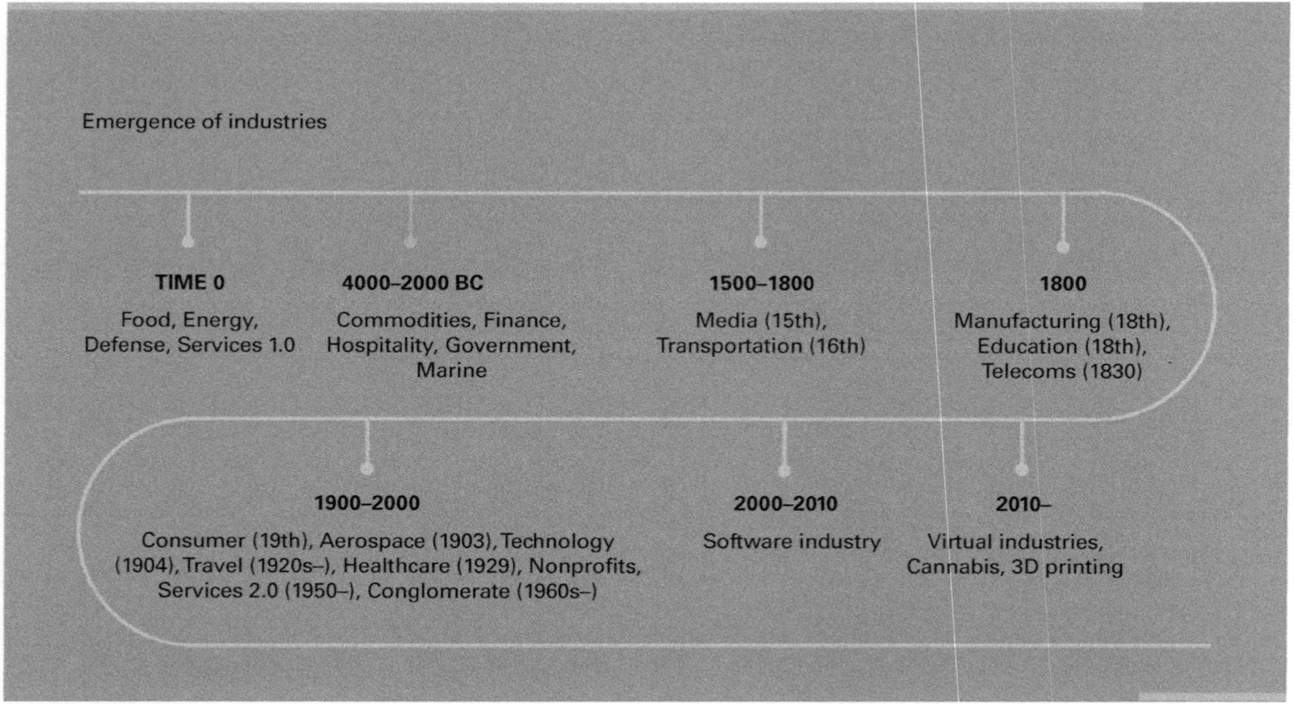

Emergence of industries

TIME 0

Food, Energy,
Defense, Services 1.0

4000–2000 BC

Commodities, Finance,
Hospitality, Government,
Marine

1500–1800

Media (15th),
Transportation (16th)

1800

Manufacturing (18th),
Education (18th),
Telecoms (1830)

1900–2000

Consumer (19th), Aerospace (1903), Technology
(1904), Travel (1920s–), Healthcare (1929), Nonprofits,
Services 2.0 (1950–), Conglomerate (1960s–)

2000–2010

Software industry

2010–

Virtual industries,
Cannabis, 3D printing

Amazon will fail. Amazon will go bankrupt. If you look at large companies, their lifespans tend to be 30-plus years, not a hundred-plus years" (Steinbuch, 2018).

Each industry gradually acquired its own professional societies, forging strong internal bonding, and set professional identities, concerns, and even a class consciousness (Marx, 1848) of "us" vs. "them." In the 21st century, however, many of the distinctive features of each industry flow into one another and the concept of industry is increasingly less useful as a means to distinguish professional concerns, patterns, and trends. That issue notwithstanding, Figure 4.2 shows a shortlist of some relevant broad industries that still are reasonably distinct.

Once I developed the taxonomy and started using it, I discovered that as useful as it seemed at first sight, it was incredibly incomplete. Literally dozens of exceptions started showing up and subcategories that were more than smaller markets pushed themselves to the forefront. The reality is that the distinction between an industry, a market segment, a sector, and an emerging focus area is not so large. After all, what does it mean to call something an industry or a market?

The traditional meaning of a market is a physical market and the layout is perfectly obvious, apart from the ownership structures which may still be quite opaque. However, these days, the moment you can delineate a set of actors and can count the amount of investment happening in a field, you can claim it is a market. Nevertheless, that does not mean that all such smaller market areas necessarily operate as a market. The challenge is that in today's marketplace, markets may not even have the same view of themselves. For example, the financial services firm Fidelity arguably now thinks of itself as an IT company. How does that fit this model? It clearly does not.

Advances in technology have helped a range of industries to move forward, achieve efficiencies, even transform their subject matter, perhaps notably manufacturing in the second industrial revolution of the last century and now in the fourth industrial revolution of the past few decades. As most of us know, *digitalization more generally is becoming a backbone of all industries* yet is having different effects in each industry. Sometimes technologies even foster new industries, which is about to happen with 3D printing, which is on the brink of fostering a distributed production line that will reshape not just manufacturing but also consumption, perhaps blurring the boundaries between the two forever.

Similarly, as mobility-as-a-service, AI, and autonomous driving take hold, the transportation industry is about to be drastically reshaped. The types of skills needed to succeed, and the type of infrastructure future transportation modes will rely on, are becoming drastically different from those required in the past. However, again, it is the interplay of tech innovation, business models, regulation, and social dynamics that will determine the outcome, not one of them alone.

The way industries morph into each other, change and interact, and evolve is not easy to observe in real time. Tracking industries is not as easy as just following macroeconomic variables year to year and corporate fiscal quarter by quarter. It cannot be induced from tracking individual companies either. Rather, the core argument in my book is that a complex interplay of innovation, windows of opportunity, government regulation, and social dynamics creates and sustains them. Take your eye off of one force of disruption for an instant and that's perhaps when the change happens if you don't watch out.

Time and time again, even those corporations with the biggest R&D departments have been hit by surprises. Note that the telco industry has among the world's biggest research teams and still was, largely, unable to foresee the internet, a technology directly disruptive to its other cables, switches, and networks, and eventually forced some of them into the content industry, which they all were ideologically against as network providers.

Becoming aware of how technology impacts new and old business models

A business model describes how an organization creates, delivers, and captures value (Osterwalder and Pigneur, 2010). As such it includes all aspects of a company's approach to developing a profitable offering as well as the method by which that offering is delivered to its target customers. In fact, Sinfield et al. (2011) claim these six questions need to be answered: Who is the target customer? What need is met for the customer? What offering will we provide to address that need? How does the customer gain access to that offering? What role will our business play in providing the offering? How will our business earn a profit?

Nailing the four traditional business models

There are four traditional business models: manufacturer, distributor, retailer, and franchise, and a myriad of others that mostly are offshoots or blends. Manufacturers produce their own products and typically sell them either wholesale direct or through a distributor. Distributors don't produce their own products but sell other people's products as a third party—having said that, their channel can be a hyper-innovative product platform such as is the case with Amazon's website. Retailers have a storefront and these days also an e-storefront where they advertise and sell other people's products at a markup. Franchises let franchise owners purchase the right to run a highly standardized operation from their own location—typically providing a mostly mandatory global marketing package and a supply chain with preferred vendors, and in return the profits are shared after a precise formula.

There's nothing wrong with any of the traditional business models. In fact, there's probably significant innovation mileage left in each model. Experience retail will continue to evolve as long as humans don't fully evolve into a virtual species (we are far from it).

Learn from previous work on business models

Luckily, we do not have to start from scratch. Various strategic frameworks have been proposed to go about learning, testing, or grasping business model assumptions. I have listed just five of them, in the order in which they appeared in print—although some of these ideas were circulating way before.

Business model books to learn from include *The Business Model Navigator* (Gassmann et al., 2004), *Blue Ocean Strategy* (Kim and Mauborgne, 2005), *Change by Design* (Brown, 2009), *Business Model Generation* (Osterwalder and Pigneur, 2010), and *The Lean Startup* (Ries, 2011). The canvas approach (Osterwalder and Pigneur, 2010) is quite practical and encourages radical experimentation. Design thinking (Brown, 2009) is a paradigm for how to approach co-developing products with clients in mind from the get-go. Blue ocean (Kim and Mauborgne, 2005) is a much-heralded tactical approach to finding uncontested space instead of being stuck in highly competitive markets. Lean startup (Ries, 2011), which arose out of Eric Ries' own experience with product failure, preaches to quickly build a minimum viable product (MVP) and then conduct tests with customers through interviewing them about their experience and iterating quickly to improve through feedback.

However, it is important to realize that disruptive innovation occurs through inventing something new that customers may not have the imagination to perceive when presented with an early prototype, Steve Jobs' work at Apple being the best example. At the end of the day, putting power back in the hands of founders is also important.

Each approach has its own merits: simplicity, empirical basis, personal touch, case study examples, or novelty. None of them has all the merits in equal measure, for instance a careless adoption of either the lean startup method, business canvassing, or even design thinking as such could lead to false negatives, meaning good ideas are mistakenly rejected based on a few random customers' bad feedback. Similarly, Blue Ocean is touted as a way to reject the either-or of value innovation or low cost, although the multiple choice questions the authors tout as essential to ask yourself (in terms of usefulness, cost, price, and adoption) are a highly simplistic view of disruptive innovation and only a source of temporary competitive advantage since others may well catch up and could deploy the same spin.

Overall, frameworks have a place. However, it is important not to get *stuck* in one or a few such conceptual worlds, because the world changes faster than frameworks do. I would like you to apply that same type of skepticism to my current book, which, of course, builds on earlier thinking, including the PEST framework (Hague, 2019) as well as a host of other traditions which I am fairly transparent about.

There is nothing magical about frameworks and they need to evolve over time. If their empirical basis ceases to be valid, they should be reworked or abandoned. Herein lies the danger of old mental models—they can become as much of a problem as an opportunity to simplify things. The main value of frameworks is their simplicity and shorthand, which may become more important in an increasingly complex world. Conversely, if you simplify a phenomenon that in reality is more complex, you are shooting yourself in the foot and would have been better off dealing with that complexity head on.

Spotting emerging business models—a challenging task

In contrast, newer, disruptive business models are relatively poorly understood and are mostly studied in the context of new technology (Schiavi and Behr, 2018). As history teaches us, the novel business model of today may not help us understand what will work tomorrow.

Looking at business models in a more complex setting, taking into account technology, regulation, and social dynamics, is rarely achieved, unless you are the disruptor, an entrepreneur tracking a variety of disruptions to capitalize on them, a consultant selling change-related services, or perhaps the party being disrupted. Studying emerging business models is particularly fraught with pitfalls, as the awareness that a business model is becoming successful is typically possible only after the fact. There is a creative aspect to business models that is often missed. The entrepreneur must make decisions based on little true information; it is often more of a hunch about what might work, followed by testing and experimentation to see whether the hunch pans out.

If startups don't find an initial market based on the business model they chose originally, they typically try to pivot, which means they try out other business models or markets. This type of business model experimentation is easier for startups but can also be performed by larger entities, especially in connection with new product launches.

The relationship between business models and technology

There are many lists of business models found online for us to explore (Board of Innovation, 2020). However, they each fall into a small set of main categories. Right now, I would say there are nine tech-enabled business models (as illustrated in Figure 4.3), although the situation is fluid and evolving. As tech, idea, and consumer demand changes, there is always the

Figure 4.3 Tech-enabled business models

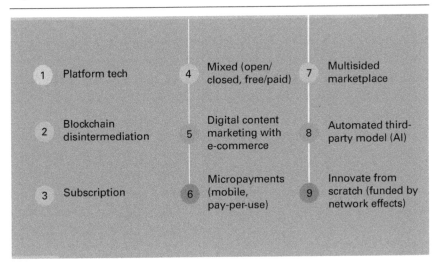

possibility that a new, clever business model will emerge. I will quickly outline what they are all about.

Platform tech—last decade's dominant approach

Platform technologies are those which build on network effects created either by a previous tech platform which one piggybacks upon or by a firm's own platform, and these are a staple of internet-based business. Many books have been written about them (McAfee and Brynjolfsson, 2017; Parker et al., 2016). Examples often used are ecommerce platforms (Amazon, Shopify, WooCommerce), mobile ecosystems (iTunes, Android), music streaming (Pandora, Spotify), social networks (Facebook, Instagram, TikTok), or the ever-present Amazon, which is a combination of e-commerce, cloud computing, digital streaming, and artificial intelligence, plus perhaps a little bit of a social network, given its emphasis on product reviews as well as the large third-party community it enables and facilitates. What's interesting is that the platform ecosystem is highly dynamic. There is no guarantee that a successful platform one year (or month) will be the successful platform the next, although economic forces allow these platforms to sustain themselves by incorporating elements from other platforms the moment they become perceived as outdated.

Blockchain intermediation

Blockchain (see Chapter 6) is a highly flexible platform technology that allows for a new set of peer-to-peer-based business models. *Peer-to-peer* refers to the fact that there is no need for a middleman, the transactions are facilitated by an open book of exchanges to which, in theory, every actor in the network has access and which helps create a trusted network.

Blockchain-based business models use tokens to make profits. The first example would be models based on utility tokens, the most direct example of which is when they are used as a currency, notably in cryptocurrencies such as Bitcoin, Ripple, and Ethereum. However, tokens in the blockchain ecosystem can be used for many other types of value-based exchange that don't involve a direct relationship to traditional money, for instance tokens can give access to rights and responsibilities, and value inside the ecosystem (but not valid outside it—such as play money inside an arcade game facility).

Security tokens are investment contracts which, for instance, allow you fractional ownership and trading of real assets. Lastly, the rapidly emerging blockchain development platforms, including BigchainDB, Ethereum, Hyperledger Fabric, R3 Corda, Ripple, Cardano, and EOS, provide new functionality that will allow continued innovation of the platform itself, extending its use cases and reach to new industries and sectors.

The emerging use case for blockchain goes far beyond the financial industry

The most talked-about use case for now is within financial services, where the fintech community's smart contract platforms would drastically change the way contracts are made, maintained, and used as a business tool. Notably, it would enable the user and creator of data to be in control. Today, once you give away your data, whether it is to the government or to an e-commerce provider, you essentially give away the right to hold that data indefinitely. There is little chance you can revoke the right, or even if you do, they still have the aggregated data about you. The early player in this field is Ethereum, which combines a computing platform, operating system, and smart contract functionality to run decentralized digital applications, or "dapps." At the moment, there are still scalability and other technical issues with the smart contracts-based solutions, which so far has hampered their widespread use. This may change rapidly.

Blockchain-as-a-service (BaaS) is a newer way to allow non-experts to deploy blockchain's effects without getting into the nitty-gritty of the technology itself, akin to the way cloud-based web-hosting services facilitate running a website. This approach is likely a precursor of things to come, where blockchain could become a dominant platform by which we conduct business across all of industry, taking on the same basic role that money plays now, but with much more attractive peer-to-peer dynamics which cut out the middleman without resorting to the limitations of one-to-one barter, which we know from the basic physical marketplaces of the past.

Mixed business models—open source and beyond

There is a plethora of mixed business models these days. Basically, a mixed model combines aspects of several business models, for instance combining

B2C and B2B sales on a single website, combining social and business goals, or indeed even a mix of three or four models, such as being a manufacturer, a distributor, an e-tailer (Walmart.com), and a retailer (Walmart) at the same time, combining private labels (319 private label brands in 20 categories) and white labels (few left now), even experimenting with subscriptions (Walmart+).

Mixed business models are mixing two separate but related concepts: free vs. paid and open vs. closed. Free as in free beer is relatively straightforward, paid as in paid beer in a bar equally so. Open is slightly different and refers to source code but also, as indicated above, to openness of the process and approach to collaboration itself. Closed specifically refers to the licensing model being proprietary, that is, ownable by a business entity. The related term of *interoperability*, which we explored in Chapter 2 as we discussed standardization, is an important gateway concept. Interoperability has the mixed blessing of being the go-between of free and paid, closed and open, in that it allows communication, access, and collaboration through interfaces that are commonly understood and described. Mixed business models tend to need interoperability to work well. Without interoperability, there is lock-in, that is, you are confined to whatever technology a vendor supplies you with, because others cannot access the specifications to work together with it.

Many mixed business models are based on the razor and blades principle in which one item (e.g. the razor) is sold at a low price in order to increase sales of a complementary good, such as consumable supplies (e.g. the blades). The principle is at play in products as different as the printer market (printers and ink cartridges and now 3D printers and print materials), computer games (gaming consoles and games), and any kind of time-limited free trial of both physical and virtual goods. A variety of this approach is product tying and bundling, such as when a cable provider supplies you program packages from content producers (TV channels) that contain only one channel you want and dozens you don't want but cannot opt out of, or Amazon's Kindle e-reader and its e-books in the proprietary Kindle format called Mobi. The business risk with mixed models is that customers can feel tricked, nickeled, and dimed if they don't fully understand what they are signing up to and might cancel once they realize what the deal is.

The discussion needs to start with open source, a type of computer software in which source code is released under a license in which the copyright holder grants users the rights to use, inspect, modify, enhance, and distribute the software to anyone and for any purpose. This type of code was born out of academic norms for open exchange, collaborative participation, rapid prototyping, transparency, meritocracy, and community-oriented development, which historically were prevalent in university computer departments but which are now much more widespread across departments and sectors of society.

Open source software has been used at least since the Open Source Initiative was formed in 1998. However, it was not until the early 2000s that dual licensing emerged as an open source business model, and it typically refers to the release of a software component under two licenses simultaneously: a traditional proprietary license and an open source one. There are also several different open source licenses, starting with the copyleft license and onto more permissive licenses such as the BSD, MIT, or Apache licenses which allow commercial implementations as long as copyright is acknowledged.

Crowdsourcing, which typically refers to the practice of soliciting services, ideas, or content from a large group of people online, was coined by Jeff Howe in a 2006 article in *Wired*. Its business model analogue, crowdfunding, the practice of funding a project or venture by raising small amounts of money from a large number of people, was coined by Michael Sullivan, an entrepreneur looking for backers to help fund his video-blog project, in 2006.

Freemium is another linked concept where the product or service is partially or initially free but prices increase after a period of time or as the user realizes they need additional functionality to carry out the business they signed up to accomplish.

Digital content marketing with e-commerce

Content marketing is a thriving business model nowadays, which is interesting because some traditional content businesses (e.g. newspapers) have needed to find new business models for their content. Newspapers, of course, rarely made a living from the content itself but were advertising based. However, they saw themselves differently, thinking of themselves mainly as providers of content (at least, the journalists did, perhaps not the owners). However, content marketing is different because the content is explicitly not the main product of the business providing it—it is the add-on used to create goodwill. The bulk of content marketing is free or freemium and might be undertaken purely to build brand awareness or loyalty as a first step toward value-based outreach.

E-commerce, particularly through online marketplaces, has similarly taken off over the past decade, as the traditional retail model suffers from physical limitations, inventory risk, congested retail locations, and other challenges.

Mobile transactions—micropayments, pay-per-use

The definition of a micropayment varies. Some companies recognize transactions below a dollar as micropayments, others classify micropayments as amounts below $5.00, below $10.00 (like PayPal), or below $20.00. If you look at your bills, you'll realize the percentage fee of micropayments sometimes can be up to 40 percent, as is the case in the freelancer platform Fiverr, which offers services advertised at $5 and charges a $2 processing fee. Micropayments comes in many forms, from prepay (seen in social media sites, online games, and newspapers) to post-pay (online music sales) to collaborative to pay-as-you-go (prevalent among software as a service providers), and each has different commercial implications, of which the consumer is not always consciously aware.

Beyond enabling all other business models, mobile transactions have also created their own. The best example is how micropayments have enabled pay-per-use at a much more granular level. Pay-per-use refers to a business model that charges per unit used, either over time or by quantity. That, in turn, fosters new types of behavior. We are no longer so worried about our "card being swiped" for microtransactions through PayPal because the perception is that the fees are that much smaller than in the credit card era of the 2000s, even though that is not necessarily true.

In reality, it is the digital delivery that makes the biggest difference because delivery costs are not an issue.

Multisided marketplaces

Two-sided marketplaces facilitating direct interaction between suppliers and customers became prevalent throughout the past decade. Many models exist, including matchmaking and brokering, commissions and auctions. Early examples included videogame platforms, such as Atari (1971), Nintendo (1972), Sega (1965), Sony PlayStation (1994), and later Microsoft Xbox (2001), which, as described by Rochet and Tirole (2005), need gamers to attract designers to develop games and need games for their consoles to attract the gamers who will buy and use their consoles.

Uber (transportation, food), eBay (used goods), and Airbnb (lodging) are all multisided markets. Other examples include Apple, Microsoft, Google, Alibaba, Amazon, and Facebook, which are also called *platforms* because they are tech-enabled marketplaces.

Uber owns no vehicles and does not make food. Facebook creates no content. eBay and Alibaba have no inventory. Airbnb owns no real estate. Multisided markets bring together two or more interdependent groups which need each other in some way and then provide a multitude of ways in which they can interact directly, where each transaction typically is monetizable.

To achieve such a position, entrepreneurs identify markets where there are many, displaced actors with ongoing coordination problems or when there is a strong perceived benefit to facilitating transactions (even if those are already happening through another channel). Second, entrepreneurs need to overcome the chicken-and-egg problem of how to create demand simultaneously on several sides of the platform. Typically, this is best overcome if one side, typically with more elastic demand, is subsidizing the other, which is harder to achieve so that the customer acquisition process and lifetime value are optimized. In fact, what characterizes multisided markets is their price structure. All of this assumes that the overall venture still is (or will become) profitable based on economies of scale. Moreover, the entrepreneur needs to have the resources and time to let these dynamics play out (Parker et al., 2016).

When they succeed, multisided marketplaces create what economists call demand-side economies of scale, more commonly known as network effects. These result when the value of a product or service changes in a positive way as more people use it and especially if that value rises exponentially, such as how the value of having a cell phone increases when more people have cell phones.

Automated third-party models

AI is arguably becoming the most important general-purpose technology in the world, although so far it acts on only small parts of the problem sets that go into a business challenge. As AI rapidly commoditizes, automated third-party models have started to emerge, notably based on the controversial field of AI called image recognition. In 2019, Dreamstime, the world's largest community in stock photography, released its product PhotoEye AI, designed to help companies find, filter, and edit content with innovative assessment and optimization tools. Various forms of due diligence are also now being offered through AI-based models accessible through third parties in the emerging third-party risk management (TPRM) market.

Exploiting the lack of a formal Chinese credit system, the Chinese firm Ant Financial, spun out of Alibaba, conducts consumer lending, money market funds, wealth management, health insurance, and credit-rating services using machine learning to analyze large amounts of data to prevent attempted fraud in cashing loans and buying medical insurance with hidden diseases. Ant Financial is about to create a private social credit system fueled by AI. The scale, scope, and learning speed of its system is astounding, as it is already serving one billion clients (Iansiti and Lakhani, 2020).

The overall logic at play is to make AI available to firms that are not AI experts so they can build their business model around it. That way, AI is starting to rewrite traditional and even newer and emergent business models. AI is slowly starting to question the very meaning of ideas, innovation, and inventions that previously were the territory of human intellect, such as creative works, advanced manufacturing decisions, intellectual property rights, driving, medical diagnostics, and much, much more. With the provision of synthetic data, e.g. data that is explicitly created for the purpose of testing new AI methods but doesn't rely on collected data, or even federated models where the entire model is trained by a third party so you can immediately put it to use, people are starting to question the necessity of deep in-house capability in many fields. These business models are enabled by the intersection of on-device AI, blockchain, edge computing, and the Internet of Things (Leopold, 2019).

Overall, AI-based business models will first emerge through the applications that have the longest track record of radical innovation through AI, such as face recognition, sensing, advertising, and product targeting, which might mean that the industries that stand to benefit the most initially will be marketing, automotive, supply chains, and manufacturing. The greatest potential, however, must surely be in healthcare, that gigantic sector with so many unsolved, multibillion-dollar problems waiting to be underpinned by an AI-enabled business model. Moreover, any sector that has experienced slow digitalization is going to see the biggest gains through AI because those sectors stand the chance to leapfrog the 30-year digitalization boom enabled by the internet into something instantly more magnanimous.

Innovate from scratch

In this business model, which is one startups deploy and corporations can emulate only partially, the value differential may be created entirely by the network effects, or there can be other inherent disruptive advantages of the business model which does not rely on such effects. Such is the case with new product innovation where the brand is the differentiator.

The term "from scratch" is a bit of a misnomer, since nothing comes from nothing. There is always a context within which innovation happens, even disruptive innovation. Similarly, even disruptors stand on the shoulders of giants since their innovation is likely to be in only one aspect and might rely heavily on others for the total product delivery to work.

Mobility-as-a-service, the move away from personally owned modes of transportation, is one such emerging business model, which essentially puts such a spin on transportation by moving it away from supplying the infrastructure, its most distinct and nagging characteristic which has held back innovation. Uber is in fact a multisided business model and its rise was based on innovate-from-scratch.

How startups upend markets

Occasionally, a tech startup, or more typically a cluster of tech startups with similarly radical technology enhancements, will upend entire markets, or at least a market segment. In recent memory, it has happened with many unicorn startups, lately particularly with software platform startups such as Intel, Microsoft, SAP, Thomson Reuters, but also in sectors as diverse as transportation (Uber) and hospitality (Airbnb).

Business models that are almost the same as older ones, or which seem like they have been tried before, are among the most exciting. They are also satisfying when they work because they tend to disprove venture capitalists who skew toward disruptive innovation as opposed to incremental innovation, even if both can have monumental impact. For example, Warby Parker's new take on being a middleman in the market for eyeglasses were turned down by several investors back in 2010. However, Warby Parker had spotted a near monopoly by Luxottica in designer eyewear, which had led to artificially inflated prices. Warby Parker now operates 90 physical stores in the US and Canada and is reportedly valued at $1.7 billion. You heard it right, that's a startup with a successful brick-and-mortar retail strategy (pre-coronavirus).

Subscription—a business model based on "the forever transaction"

The *forever transaction* is another key idea that has emerged (Baxter, 2020). It builds on the basic principle familiar to all mom-and-pop neighborhood businesses: get to know your customers well and stick around to build a long, ongoing relationship. Startups are not alone in deploying this approach, but they are uniquely suited to succeed with it. Dollar Shave Club, which sells razors, was so successful that Unilever bought the company in 2016 for $1 billion.

Powerful brands like Nike, Spotify, LinkedIn, Netflix, Amazon, and Target all build a membership platform thinking about their customers in the very long term. Other, smaller, and emerging companies attempt to do the same thing, for obvious reasons. However, the downside is a tenuous trust relationship that also could be severed. When broken, that forever transaction becomes like a divorce, it may lead to aggressive behavior to prove you are different. This is when consumers switch from Coke to Pepsi, from HBO to Netflix (or back), from Nike to Adidas, from Amazon to smaller retailers (many), or from Target to Walmart. The fake forever transaction is somewhat unethical in its stretch of the imagination—are you really that tied to these products and brands when it comes down to it? Unless it is a technical lock-in, the membership is all a mental creation and can be changed. Even these small, artificial barriers such as being locked in for an extra month or year are not going to cut it if a consumer truly wants to sever the tie. What companies—rightfully—bank on is that this is a relatively rare occurrence.

What is typically less often talked about is how startups often need to be connected to the distribution networks of existing massive companies in order to have the kind of leverage a platform business needs. This is what explains the mergers and acquisitions (M&A) spree that larger companies engage in, swallowing up startups that are starting to build a groundswell in a market the larger company either controls or could control with the addition of startup capability. All of big tech maintains its dominance this way (IBM, Microsoft, Cisco, Oracle, Adobe, etc.). For instance, in 2018 Adobe acquired marketing software company Marketo for $4.75 billion in order to gain leverage in the customer experience marketing market.

Subscription business models, where customers or users must pay a recurring fee, have been around for a long time—recall mail order book clubs and wine clubs and ticket sales. They have always been fantastic builders of brand loyalty because of the long-term nature of the relationship, which makes it sticky. But it is the advent of software-as-a-service (SaaS) subscriptions that changes the game. Now, the scalability and flexibility ensure global reach for digitalized information goods, whether they be magazines or other publications, as well as for physical goods, which is the true game changer. Who would have thought that subscribing to razors (Dollar Shave

Club) would have been such a hit? But it is the shift from delivering products to delivering services which represents the real change away from the 20th-century manufacturing economy.

According to Zuora (2020), the subscription economy and the sector are growing fast. Zuora cites Credit Suisse stating a whopping $420 billion was spent on subscriptions in the US in 2015. For example, Zuora is a SaaS platform that automates all subscription order-to-revenue operations in real-time for any business, providing subscription-as-a-service or basically an end-to-end subscription management solution.

What are the truly novel emerging business models?

How do you stay on top of emerging business models? The only way to know for sure is to be actively involved in experimenting with business models yourselves.

Futurist Peter Diamandis (2020) thinks there are seven emerging business models right now: the crowd economy, the free economy, the smartness economy, closed-loop economies, decentralized autonomous organizations, multiple-world models, and the transformation economy. Arguably, none of these is particularly new. Post COVID-19, many of these models will be put on hold while others take their place, notably (a) a rethinking of what it means to combine an efficient with a resilient supply chain, (b) a speedier path to the future of work (yet still a rocky one), and (c) a regression to the mean, which mustn't be underestimated (as people long to go back to their habits).

Typically, the advice would be to stay on top of emerging technologies since many new business models build on a technological breakthrough. It should be no surprise that I feel that's insufficient. You need to take into account governmental forces such as policy windows and regulatory environment. You need to be an astute student of social dynamics. Most of all, you need to be sensitive to feedback. It is not enough to experiment if you don't draw the right conclusions based on that experimentation.

If I were to speculate on a few emerging business models in the next decade, I would think of the following: online/offline integration (regardless of industry), total brands (supplying all major cradle-to-grave products for various consumer segments in a convenient manner—an extreme version of

Figure 4.4 N+1 combinations of business models

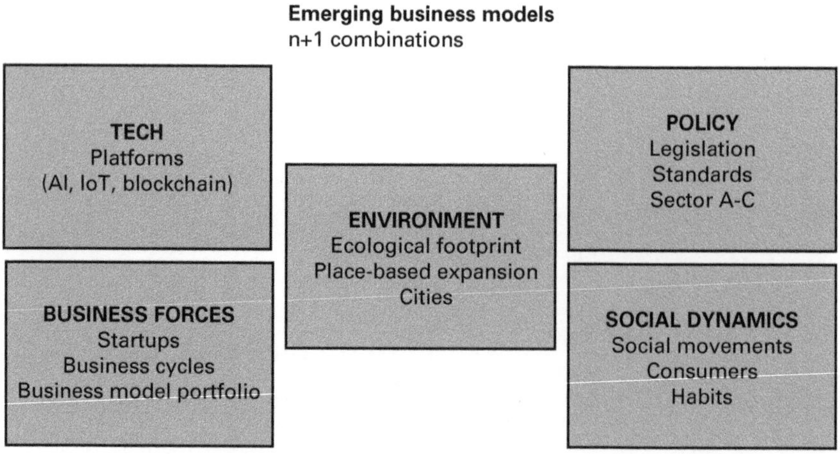

Emerging business models
n+1 combinations

TECH
Platforms
(AI, IoT, blockchain)

ENVIRONMENT
Ecological footprint
Place-based expansion
Cities

POLICY
Legislation
Standards
Sector A-C

BUSINESS FORCES
Startups
Business cycles
Business model portfolio

SOCIAL DYNAMICS
Social movements
Consumers
Habits

the forever transaction described by Baxter (2020)), and multiple-business model tie-ins including freaky combos that somehow work (e.g. both a bank and a lifestyle brand).

Beyond that, I would assume that each emerging platform technology we described in Chapter 2 will enable at least 2–5 new business models, and if we add the potential n+1 combos of each, there are hundreds new models emerging over the next decades (see Figure 4.4). Each will expand the global market, although some may be quite niche. Another approach would be to look for new business models along each of the four forces and combos within and across forces (see the exercises at the end of the chapter for examples to let your imagination run free).

For example, AI (tech) + privacy (policy) + subscription (business model) + Black Lives Matter (social forces) + COVID-19 (environment) might produce things like public data trusts, which foster trustable stewardship over public data, by clearly delineating the rights that should be invoked, advocated by the think tank Atlantic Council (2020).

Strategy frameworks—business models put into a larger context

While business models explain how a business makes money, they may not contain all the tools a business can use to plan ahead. For that, you need a strategy that considers a larger set of disruptive forces. This is why, in my mind, strategy frameworks typically extend beyond business concerns and into societal factors. The FIDEF framework that forms the basis of the book you are reading right now is an example of a strategy framework. I will briefly touch on a few others.

A strategy framework is a way to understand the business in which you operate by ways of considering the forces surrounding you. When used appropriately, a strategy framework puts you in a position to plan for the medium to long term because it helps you achieve clarity of thinking and a way to focus attention. Strategy frameworks are conceptual tolls, at times called "mental models," that frame reality and lead you to look for certain things, not others. What this means will become clear in a second.

The best recent book on strategy frameworks is *The Business Models Handbook* (Hague, 2019). Just watch out for Hague's terminology. He sketches 50 "business models," although what he writes about is what I would more accurately call strategy frameworks (e.g. the Ansoff matrix, the Boston matrix, Porter's five forces, Kano's customer requirements) because they are better at analyzing businesses and acting on it (strategy) than implementing new ones (deploying a business model).

Hague usefully opines on the use cases for each of the 50 frameworks, putting them in six categories: marketing, general business strategy, pricing, innovation, product management, and customer analysis. Interestingly, the fact that many frameworks originated in the field of marketing should be a warning bell because it illustrates to me that the framework might not be as innocent as we assume; it is itself part of a marketing ploy on behalf of the author of the framework. Most frameworks of this kind were introduced by individuals connected to consulting firms or business schools. Also, frameworks designed for marketing cannot be used to analyze a different phenomenon without being adapted. I have seen frameworks used in radically differently ways than those for which they were intended. That may work, but the rationale has to be clear. Is the framework still predictive of a situation on the ground? Using a bad or outdated framework only clouds thinking.

Figure 4.5 Strategy frameworks

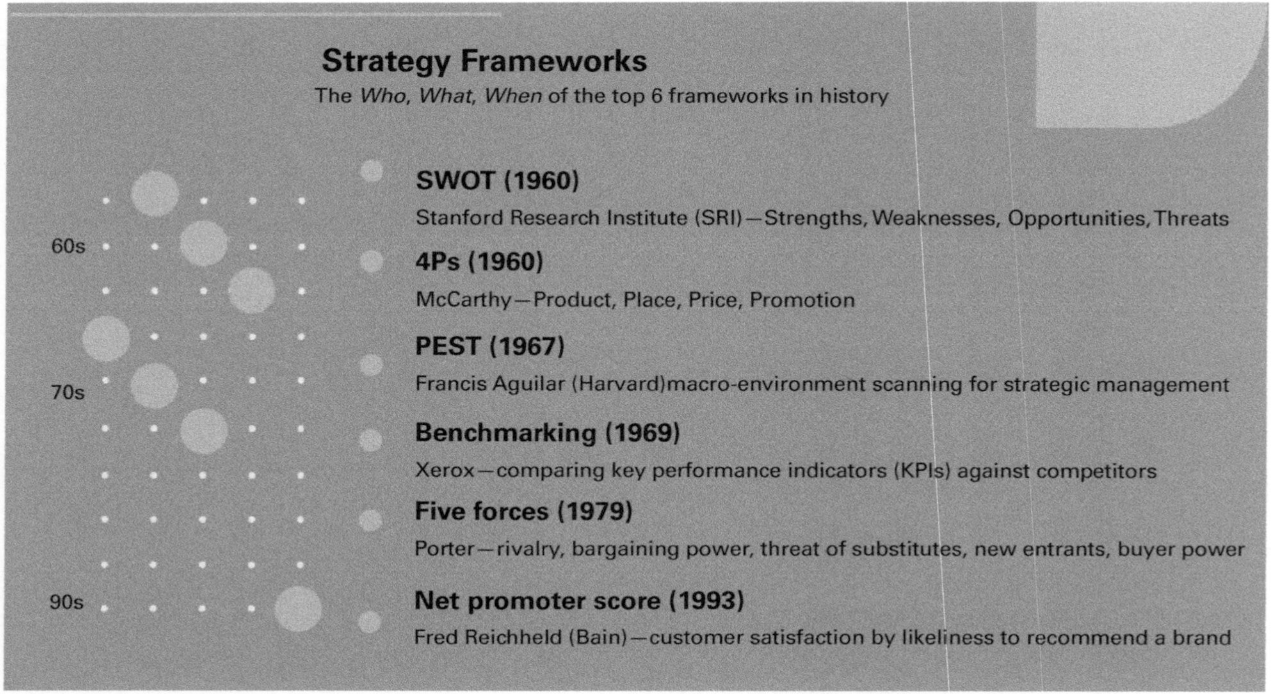

Strategy Frameworks

The *Who, What, When* of the top 6 frameworks in history

SWOT (1960)

Stanford Research Institute (SRI) — Strengths, Weaknesses, Opportunities, Threats

4Ps (1960)

McCarthy — Product, Place, Price, Promotion

PEST (1967)

Francis Aguilar (Harvard) macro-environment scanning for strategic management

Benchmarking (1969)

Xerox — comparing key performance indicators (KPIs) against competitors

Five forces (1979)

Porter — rivalry, bargaining power, threat of substitutes, new entrants, buyer power

Net promoter score (1993)

Fred Reichheld (Bain) — customer satisfaction by likeliness to recommend a brand

60s

70s

90s

While I agree that there are at least 50 relevant strategy frameworks out there today, I'd say only 5–10 of them have had the kind of pan-industrial impact that you need to worry about in terms of capturing how business models cause disruption. The top five are SWOT, 4Ps, benchmarking, five forces, net promoter score and PEST (see Figure 4.5).

SWOT—overly simplified

Considering strengths, weaknesses, opportunities, or threats (SWOT) is the simplest way to think about any situation that might arise or to ponder your options in business or indeed in life, and deserves a mention even as a generic strategy humans tend to resort to. The drawback is that this is not really a framework as much as it is a set of four terms for which you have to entirely figure out the meaning on your own. The distinction between weaknesses (internal) and threats (external) is slightly unnecessary, likewise the distinction between strengths (internal) and opportunities (internal and external). That means you are really left with two recommendations: study your strengths and weaknesses and take a look at the external environment.

The 4Ps—price and promotion are the hard parts

Grasping the importance of the 4Ps of product, place, price, and promotion is essential to understanding the basics of marketing, a discipline which underpins our consumer society. Specifically thinking of how to build brand loyalty, which has increased in importance since the framework came out because of greater competition and complexity in the market, Londre (2007) includes a total of nine Ps: planning, product, people, price, promotion, place, partners, presentation, and passion. However, as you extend a framework it rapidly loses its simplicity and as a result its usefulness declines. The most difficult factor to figure out is pricing vs. promotion, and that field has significantly evolved with the rise of machine learning applied to a myriad of factors as well as sales data.

PEST—macro-environment scanning for strategic management

When Francis Aguilar (1967), a Harvard professor, issued his book *Scanning the Business Environment*, he spawned a host of followers who have tweaked his original four factors—Economic, Technical, Political, and Social influences—in many directions. My own FITEF framework stands on the shoulders of that work, too. The challenge over these past 50 years has been the lack of rigor around how to use it, which factors go into it, and most critically, which subfactors are most important. The answer depends on what you are using it for, the data available, but also on your systemic understanding of society. Overall, PEST is mostly used to get a quick take on the external environment that might influence a decision, a business, or indeed any phenomenon. In contrast, in my own book I have rigorously identified the most salient subfactors and tried to derive cognitive strategies as well as provide exercises that hone the way this type of framework is used so it can respond to the 21st century's challenge of managing technology.

Benchmarking—compare your approach to others and learn from it

Benchmarking is also so generically useful as a concept that you almost doubt it originated in 1971 around Xerox Corporation's work on quality. The concept of course was not, because it builds on *statistics*, a term that originated in the English language in 1791 and which became mainstream in the 19th century as a tool to describe large groups, populations, or moral actions, first in the military and later in cities as they grew and encountered coordination challenges. By comparing yourself or an object of study to another relevant entity you gain clarity about what you have available, about relative positioning toward other entities around you. As a result, you can make more informed choices about how to proceed. All of this is useful only if you have the right data for such comparisons and it can be collected and analyzed without too much pain.

The five forces of competitive advantage—reactive and outdated

Five forces refer to rivalry, bargaining power, threat of substitutes, new en- trants, and buyer power. The framework, which was developed by Harvard professor Michael Porter in 1969, almost instantly became the mantra for strategic analysis of competition and has influenced scores of businesses until this day. It has led to taking care in picking the industry you enter, which lines up with the focus on value carried on by Blue Ocean and other strategies as well, with the exception that to Porter it is the competition that shapes strategy and not the other way around. Today, most tech-enabled businesses would try to provide something new and escape Porter's slightly angst-filled scenario.

Net promoter score—the customer is right

Net promoter score was conjured up in 1993 by Fred Reichheld of the strategy consulting firm Bain. His team's idea was to measure customer satisfaction by their likeliness to recommend a brand. The idea seems simple enough, makes intuitive sense, and has proven relatively straight-forward to implement in surveys, at least until the onset of survey fatigue in the second wave of digitalization over the past decade. However, again, the devil is in the details. Trying to derive insight from anything less than 100 customer responses is not even pseudoscience, it is pure speculation. Lots of net promoter scores out there are based on bad data. Having said that, if you can get clients to promote you, you have a business, so figuring out why this may not be happening is crucial for success in the short term. It may be slightly less useful in the long term and for technology businesses where the product is novel and you need to educate the market first.

As you will come to realize, the last framework I include was created in 1993, which is near 30 years ago. Even as it may appear on the scene instantly, it takes time to call a winning framework. Many ideas are floating around at this moment, but it is unclear to what extent they have explanatory staying power as well as the necessary inherent brand strength to prevail.

Conclusion

This chapter explains how disruptive business forces, notably emerging business models combined with established strategy frameworks, could be leveraged to better understand the opportunities that the future of tech-nology offers to firms and individuals operating within them. However, the lesson is that even though past experience is instructive, it is no guar-antee for the future. The reason is that disruptive technologies tend to create their own business models, sectors, and entirely new value chains. It is not even certain that the platform business model which has been so prevalent over the past decade will be so in our current decade—or certainly likely not that of the next. As I've reworked the PEST model in this book, I have tried to update its assumptions to a 21st-century context and make it still flexible enough, and simple enough, to gain considerable insight in a short amount of time. However, as I will explain in the next chapters, analysis is not enough. There is no substitute for in-depth, practical engagement with technology where you are using it to solve a problem you have. Full understanding requires participating in the solution to see what it really entails, not reading about it. The fact that many corporations are flocking to startup partnerships is evidence of this trend, which is largely positive, as long as the learning and benefit are shared between the learning parties.

Key takeaways and reflections

Reflect on your own relationship to business models. What are some examples you have watched of markets being upended?

1   Try to summarize the major business models you think matter at this moment in time (without looking back at the chapter). Then try to group them together.

2   Strategy frameworks: Which three do you find the most useful and why? Now, pick one of them and apply it to a business challenge you have right now. Then, discuss your observations with a peer (or even involve them in the thinking earlier to have a sparring partner).

3   Business functions: As forces of disruption continue to impact us, which of the traditional business functions are more resilient and will largely keep their role (if any)? Which will gradually morph into other roles or perhaps even disappear?

4   Fill in the blanks in the Figure 4.6 emerging business models exercise, using the first combinations that come to mind. Try to sketch a business on a napkin based on the concept you come up with. If they are available, discuss your findings with others around you. Try to do this after each major speech or event you attend where you are receiving new impressions that impact the way you think about the next 2–3 years.

Figure 4.6 Emerging business models exercise

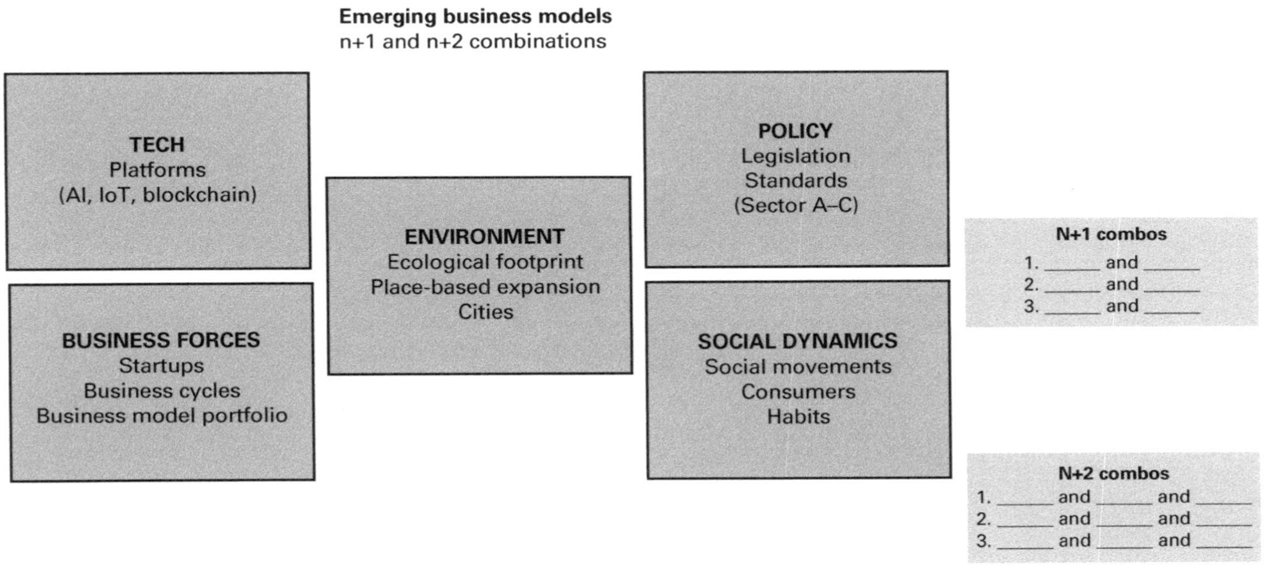

Emerging business models
n+1 and n+2 combinations
TECH
Platforms
(AI, IoT, blockchain)
BUSINESS FORCES
Startups
Business cycles
Business model portfolio
ENVIRONMENT
Ecological footprint
Place-based expansion
Cities
POLICY
Legislation
Standards
(Sector A–C)
SOCIAL DYNAMICS
Social movements
Consumers
Habits
N+1 combos
1. ______ and ______
2. ______ and ______
3. ______ and ______
N+2 combos
1. ______ and ______ and ______
2. ______ and ______ and ______
3. ______ and ______ and ______

5 Every year, at a convenient time, take stock of the new unicorn companies that year and try to describe and categorize their business models. First assemble a bit of research and then prepare an exercise for yourself (or your team) without using much more than the name and some minimal information about what each startup does. Over time, you will have a nice comparison year-over-year, and you will be way ahead of the competition in this aspect. For now, try to think of 2–3 new unicorn-style companies, what they do at the core, and ponder how they came up with their business model.

6 Spend a few minutes spelling out what business model you are currently working from. Then, imagine that this will no longer be possible, due to some calamity affecting your market. How would you pivot? Develop three ways: easy, medium, and complex.

7 Reflect on the following: It is not the number of business models you have heard of, how cleverly you can describe trends in the market to others, or anything to do with knowledge that matters: it is execution. Therefore, in thinking about business models, make sure that your thinking is sufficiently action oriented. Think, but then act. Experiment with a business model. Take a small risk. Spend time with a startup you have identified as doing something interesting. Offer to help them with something within your expertise in return for a few days of observing them at work. Set up a scheme where if you are wrong, you lose something. Challenge yourself to get better by introducing carrots and sticks to your own learning.

Social dynamics drive adoption 05

In this chapter, I emphasize the role of social dynamics in the evolution and adoption of technology trends and particularly their role in enacting, slowing, or speeding up change that stems from technology. Social dynamics can preempt, accelerate, and stall all other forces by the sheer force of numbers and by the fact that it is the combination of individual and group action that ultimately powers and sustains business, government, and technology alike.

I chart the specific influence on the future of technology from the point of view of the power of consumers, cultures, generations, and social movements, and other relevant social groups, including stakeholder groups that may represent organizational interests and/or individuals (such as associations and nonprofits). The way these groups have power is through mobilizing by representing their cultural backgrounds, by vocalizing generational experiences, and through participating in social movements voicing opinions in public.

Consumers have become immensely powerful in retail and arguably have more influence on what is being sold than shops, vendors, or even B2C entrepreneurs or product developers. Whether this will last and what the future holds for ownership, pricing, product strategy, physical retail, and a host of other issues will be discussed in this chapter. I will also consider the science behind psychographics, which was popularized by disgraced pollster/influencer Cambridge Analytica.

How a futurist thinks of social dynamics

Social dynamics qualify all other forces because they can both preempt and stall all other forces by the sheer force of numbers. People, when acting together, are what move society. On the other hand, the impact individuals can have, alone or as a group, depends on the surroundings—the groups they express themselves through, the products they have access to, the locations they frequent, the governance structure they are part of, and the business strategies they are subject to and react to.

Social dynamics, in a general sense, refers to the entire subject matter of sociology, which is the study of how human society develops, functions, and evolves. For our purposes, we are concerned with a slightly more limited set of issues: how to define social dynamics as regards the future of technology. Here's how I think about it.

Social dynamics is the sum of all forces of disruption brought about by groups that exert their influence on the development of technologies, and the products and services that result from them when enough users adopt them and find them useful in the long run.

Social groups relevant to technology can be characterized in various ways. The most important social groups for technology development in our present moment would seem to be (a) technology product developers (as a professional group), (b) the collective influence of tech startups (as a type of organizational force with a common or at least overlapping set of concerns), (c) the lobbying platform of tech platform companies (such as Apple, Google, Netflix, Uber, Amazon and Verizon, to take the US variety), (d) consumers of technology-enabled products (an important role most people play), (f) cultures (notably in the US, the EU, and Southeast Asia who jointly dominate technology production), (g) generations (cohorts of consumers defined by having had seminal sociopolitical experiences at a similar age), and (h) social movements (a type of ad-hoc group that could emerge quite suddenly based on a set of shared concerns and may enact short-term or lasting change depending on its evolution).

Because I have covered startups throughout the book and feature platform companies in Chapter 2, stakeholder engagement by lobbyists in Chapter 3, and product developers in Chapter 4, my focus in this chapter will be on understanding individual consumers (using technology for personal reasons or at work) and specific social groups: cultures, generations, and social movements (see Figure 5.1).

Figure 5.1 The subforces of social dynamics

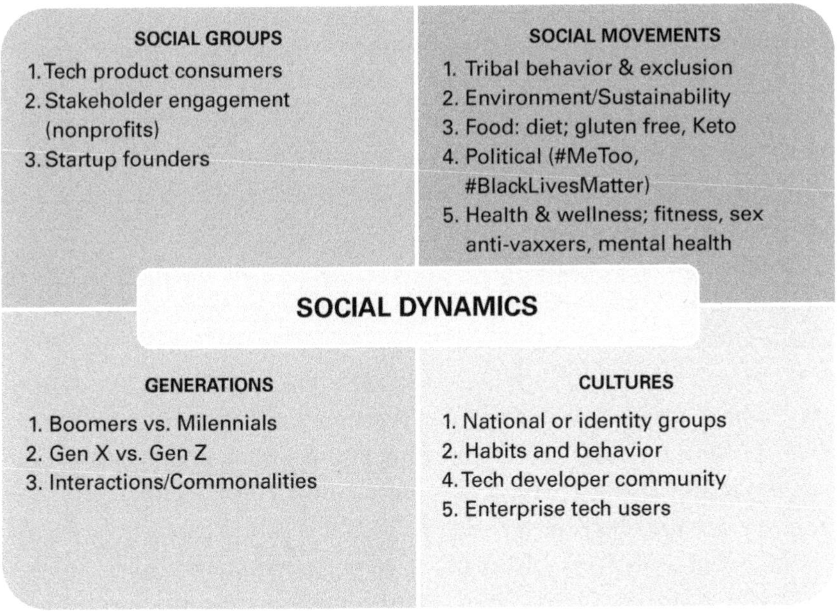

SOCIAL GROUPS	SOCIAL MOVEMENTS

1. Tech product consumers
2. Stakeholder engagement
  (nonprofits)
3. Startup founders
	1. Tribal behavior & exclusion
	2. Environment/Sustainability
	3. Food: diet; gluten free, Keto
	4. Political (#MeToo,
	  #BlackLivesMatter)
	5. Health & wellness; fitness, sex
	  anti-vaxxers, mental health

SOCIAL DYNAMICS

GENERATIONS	CULTURES
1. Boomers vs. Milennials 2. Gen X vs. Gen Z 3. Interactions/Commonalities	1. National or identity groups 2. Habits and behavior 4. Tech developer community 5. Enterprise tech users

How individual consumer habits influence technology's impact

Individual habits are more complicated than they seem. Habits are formed within social groups and social groups, in turn, influence our use of technology. Indeed, in many ways, habits are what defines a group. Great technology pioneers tap into that. Consider Apple's Steve Jobs, when he built products that not only were habit-forming but tied in with evergreen habits (listening to music, being creative, sharing content) that his company then shaped and reshaped in tune with learning more and more about these habits and how they play out in collectives such as youth, designers, and Millennials, through the product's life cycle. Companies such as Adobe have since learned to tap into the same stream of consciousness and are now much more successful because of this newly won awareness.

How habits influence our use of technology

As Nir Eyal (2016) points out in his book *Hooked: How to Build Habit-Forming Products*, apps like Facebook, Slack, Salesforce, and Snapchat need to become a habit or else they go out of business. Moreover, if the service is not in regular, frequent use, the products gradually become less useful, and eventually customers never return. Habits are among the most understood phenomena in sociology. We know that they develop out of societal behavioral patterns such as repetition and regularity, as well as being fueled by psychological needs such as safety and social needs such as belonging. But there's more. Habits form in highly specific ways depending on the circumstances you find yourself in—they have the potential both to constrain and to liberate us. Here's how.

French sociologist Pierre Bourdieu (1977) stated the highly useful point that habits, while individually held, are supported by social conditions. In fact, he maintained that habits have both an active mode (structuring behavior to consolidate change) and a passive mode (structured behavior demonstrating how we tend to operate). The passive mode has a structuring effect because of the existing structures in place which constrain us, mentally and even physically. The active mode has a structuring effect because every time we enact it, habits become reinforced. He called this phenomenon, the observation that habits are individually acted but socially constrained, habitus.

What does matter is that habits matter both on conscious and subconscious levels, and cannot be fully controlled by individuals themselves because we participate in an already created social structure. This makes the idea of a completely independent choice a fallacy. Habitus refers not only to the physical embodiment of cultural capital, to the deeply ingrained habits, skills, and dispositions that we possess due to our life experiences, but to the principle that governs repetition as a human instinct. Using a sports metaphor, it is the "feel for the game" that separates good from great and distinction from the everyday.

We are always choosing only within the constraints of the context or setting we find ourselves in, some aspects of which we would not even be able to put our finger on, unless we look more closely. At this point, an iPhone is a social construct, not just a product. Six years ago, I bought an iPhone and I've upgraded once. A year ago, I was gifted a third one, the latest upgrade. Why? Because the person who gifted me the phone wanted to give me a nice gift, of course. But also because I was counted as being (living) within the Apple IoS ecosystem (described in Chapter 2). The third time around certainly was not my choice since it was a gift. Yet, to think of the three as independent of each other would be wrong, too. I subconsciously both wanted and, on some level, needed a new iPhone. Perhaps I had even expressed that wish, although definitely not many times.

Between the gift and the gift giver there exists a field of constraints that have to do with path dependency (of the iOS ecosystem), economic status (an iPhone was still attainable), and preference (I had expressed a preference for iPhone over various Samsung models which belong to the Android ecosystem which are also in use in my family). I might have bought the first iPhone despite my habitus (although I would attribute it to user preference plus perhaps a path dependency from iPad to iPhone), but I definitely got the second iPhone because of it.

Consumer power

Consumers have become immensely powerful in retail, and it is not inconceivable to claim that today consumers themselves have more influence over what is being sold than shops, vendors, or even B2C entrepreneurs.

Consumers have so much more influence over what and how products are produced today because most products are beta tested with consumer panels or focus groups way before ever being introduced to the market, and if the response is not overwhelmingly positive, the product is shelved. Products are adjusted mid-cycle to allow for improvements. Also, the product development cycle has sped up, with new, incremental product releases often every year in some market segments, such as cell phones.

The failure of enterprise technology

Products that are not designed with the end user in mind (and their habits) won't work. However, tech products were not always sold with a specific social group of users clearly in mind, which was not sustainable and led to a lot of product failures. There is a category of products which for decades was designed to serve the needs of enterprise that purchased the product rather than those of the end user.

I am talking about enterprise software and a host of other products sold directly to businesses in bulk licenses but (obviously) intended for individual use within those companies. For example, Blackberry, Microsoft Office 365, SharePoint, LinkedIn, and many "enterprise SaaS" providers in the 1990s and early 2000s, such as Microsoft, SAP, Oracle, and Salesforce. In 2019, the enterprise SaaS market hit $100 billion, according to Synergy Research (Miller, 2019), so it is not a small market.

What's striking about the notion of "enterprise software" is that it insinuates that individuals are not the same when they use technology at work. To be effective, products that consumers use—whether they are at work or at home—need to be designed with the user in mind. The products should not be developed only according to specs presented by enterprise buyers, such as complexity, security, and failover requirements, but should have the same (or better) usability as products sold to individuals.

The rise of BYOD (bring your own device) products is the first example of trying to fix the notion of enterprise software. Beyond that, the fact that most workers blend their home and work devices when they are working from home, with or without an official BYOD policy, is further evidence of this trend. At the end of the day, we are consumers all day long and our preferences cannot be shut down when we work.

Cloud computing and subscription-based business models changed the way many of these enterprise products were adopted, notably where individuals might already have their own personal accounts on these types of software way before their employer took a corporate license, and this was way overdue.

In fact, compared with the growth of consumer tech (sometimes with the same products) in the same time period, enterprise software has been a colossal failure. Arguably, the quality of the software suite of an independent professional far outstrips that of a corporate employee. Taking into account the corporate employee's private software accounts, the picture is more mixed since most people will compensate for their corporation's shortcomings by taking privately expensed (or free/freemium) solutions on their private budget in order to stay competitive.

Consumerization of enterprise software is now slowly underway, partly fueled by the increasing army of freelancers, and will largely be a good thing for all involved, vendors, enterprises, and employees. For example, it allows self-onboarding instead of the dreaded introductory courses that corporations used to run to explain all their (crappy) proprietary software with endless plugins and quirks. The entire concept of self-service, as well as "no-code" of course, in turn challenges the very notion of why a corporation is needed in the first place, which is one of the challenges of the impending future of work set to wash over us in the coming decade. If the corporation is no longer needed to structure advanced work, why is it needed at all? The traditional answer is because it can scale products. But the benefits of that scale (to social groups) is questionable. Pay-as-you-go business models (described in Chapter 4) change the playing field by lowering barriers to entry and the sticker shock of enterprise pricing.

The impact of social groups on technology

Three types of social groups have a strong impact on the future of technology: cultures, generations, and social movements. Cultures are relatively stable over time. Generations emerge and change with the times they find themselves in. Social movements form based on individuals who develop a common sense that their voice is not being heard, taking action to change this predicament, and are more volatile social groups that don't last as long as the other two.

How a group defines and constrains individual expression of technology needs

Studying (and becoming increasingly aware of) social dynamics is not just about capturing the continuous change in consumer demands, it is also about attempting to capture and comment on deeply rooted structural constraints. The similarities that unite a social group also provide constraints. Those constraints are indeed relevant to technology evolution.

For example, citizens living in the US today are a (somewhat divided) social group, united by taxes, territory, and broad-stroke culture. Yet, US citizens are both united and constrained by a body of laws that allows certain freedoms while at the same time limiting other types of behavior. Within the US, subgroups feel very differently, depending on their cultural and educational backgrounds, purchasing power, or consumption history with technology. They aspire to buy and use different tech products. Opportunities to take part in technology's evolution differ. Finally, their attitudes toward big tech and their brand allegiances differ. The computers available at Walmart, a supermarket, look nothing like those available at Micro Center, a specialty store, and they can do very different things.

In the same way, the ties that bind a social group can either pave the way or block the acceptance and/or evolution of technological developments.

An example from the digital space would be the pivotal role of software developers, tech startups, big tech (Facebook, Amazon, Netflix, and Google), and before that, Microsoft, IBM, and Oracle, social and digital media users, American culture, Millennials, and the free software movement (with their relentless focus on privacy).

Cultures and technology

Tech innovation occurs in a cultural context. American tech giants are different from Chinese ones, European ones are different, too. They differ in their approach to privacy, their dealings with government, their view on free speech, and in a host of other ways. In fact, if we model them on our forces of disruption graph, a bunch of differences appears.

The mere fact that some technology products are more successful in some markets than in others is sufficient reason to take culture seriously as a determinant of technological success. Clearly, there are other reasons, such as the market of origin of each company. It is easier to market to your home market, being both in physical and value-based proximity. However, we cannot conclude from the notable success of a host of Chinese big tech companies (e.g. Tencent, Alibaba, Xiaomi, ByteDance) that they mostly appeal to Chinese consumers. Because they are tapping into real consumer needs for services, cultural expression, and more effective ways to transact through the internet, their domestic success is in some cases spilling over into international success.

Consider, for a moment, the viral success worldwide of the video-sharing app TikTok throughout 2019 and 2020, with 2 billion app installations and 800 million active users worldwide, ninth in terms of social network adoption in a very short amount of time—less than three years—and 1 billion videos viewed every day (Mohsin, 2020).

Cultural factors that affect whether a technology gets adopted include effortless performance on attractive metrics (e.g. viral video), low barriers to entry and required effort, degree of social influence (being perceived as "cool" among a peer group), and a set of relevant facilitating conditions, such as relative advantage, compatibility with your lifestyle, simplicity, trialability, and observability (Rogers, 1962). Conversely, users always make technology their own, so the ability to tweak a technology to your own needs is also important.

How the generational group perspective affects technology—and how it doesn't

The generational perspective has held its relevance literally for generations. However, that importance has been sustained by market researchers with a vested interest in the concept and has been fueled by market insight departments in large corporations which seek a simple fix to focus their marketing activities and target their spending. It might be time to revise its importance. Let me be clear: What generation a consumer belongs to is less important than you think, and it is becoming less and less important as this decade proceeds. Cycles are accelerated, experiences are to a much larger degree shared between generations, and social media confuses all of us with its overwhelming amount of messaging. But let's first look at how we got to this point.

The main tenet of its argument is the following: groups of people or cohorts who share birth years and experiences as they move through time together. According to economic wave theory (Sterman, 1987), there are defining economic and social events that unite people born within a 30-year period resulting in specific political and consumer values. These people belong to a specific "era"—the Depression era, World War II, Vietnam, etc. Similarly, marketing theory defines consumers into generations or "eras" but reduces age span to around 15 years.

Generational marketing—a misguided approach

Generational marketing divides living generations into the GI generation (born between 1901 and 1926), the Silent generation (born between 1927 and 1945), Baby Boomers (born between 1946 and 1964), Generation X (born between 1965 and 1980), Generation Y/Millennials (born between 1981 and 2000), and Generation Z/Centennials (born after 2001).

Its appeal is in the easy shorthand and the ability with which we can label people we don't really understand as "others." Thinking that generations are substantially different may perhaps help a marketer to target their products. However, it obfuscates reality. The reason is that many contemporary events (COVID-19, for example) affect all ages, albeit in slightly different ways. The point is that even though such events may affect young minds in particular ways (which is still possible), it simultaneously has a significant effect on all people, regardless of their age. The real difference is pre and post COVID. That would diminish the age-specific relevance of this event in terms of how it would affect consumer choices, in a way Woodstock or even the civil rights movement in the 1960s arguably did not.

That being said, these generational marketing divides are, at times, a useful frame in which to explore how social groups easily adapt or reject specific technologies. For example, given the availability of kinds of technology those born between 1901 and 1926 were exposed to, it is easy to see why the GI generation is less tech literate than the Silent generation. And while Boomers (those born between 1946 and 1964) are catching on rapidly to things like videoconferencing and email, these tools were not available in the first 15 years of their lives.

Likewise, although Generation X grew up with an "educational" experience of the internet, the sheer explosion of the mobile internet era's "entertainment" experienced by Millennials explains why Millennials are considered "digital natives." As we consider Centennials, who are also digital natives, their experience with technology is likely more intermixed with the realism of the global challenges that persist despite new technology, given COVID-19 and other incidents.

However, although, arguably, thinking of technology adoption too much in terms of generations is not a fruitful approach, it will continue to shape the availability of technology products for some time to come. "Generations" has become a truism in product development and marketing circles. I may have to spend a decade trying to take it down, and if that's what it takes, I will. Generations may or may not form cohesive consumer sentiment, but all odds are the effects are much smaller than market research firms, which preach the gospel of generations proclaim (full disclosure: I used to work for one of those).

How social movements affect technology

A lot has been written on the way technology impacts social movements, whether they be feminism, the gay rights movement, the peace movement, civil rights, environmentalists ("Fridays For Future"), political movements (#MeToo, #BlackLivesMatter, #fake news, #QAnon), and typically as a resource mobilization tool, but far less is written on how social movements impact technology. The best example would be political movements, where political actions (creating fake news) literally reshape the social media landscape by first exploiting its openness and pervasiveness, but in a second instance leading to a regulatory purge that will alter how social media operates.

We are already seeing censorship on Twitter and Facebook (which they themselves are enacting under political pressure). TikTok being banned in the US in 2020 is another example. The social outcry around disgraced pollster firm Cambridge Analytica has altered what individuals are willing to disclose to market research firms and changed the way data is collected, which over time deteriorates the quality of survey research. COVID-19 led all the major publishing distribution platforms to censor health information or even analysis on the impacts of the virus, which meant that for a brief moment at least they ceased to operate as a neutral player in the publishing process (simply a third-party distribution function in between the publisher and the market) and started using technology (machine learning algorithms) to filter for appropriate content.

Technology enables action-at-a-distance, such as engaging meaningfully in social movements remotely, through economic support, enabling remote surveillance (through drone technology, for example), or providing ideological or even epistemic support to activists on the ground by taking part in videoconferencing or email communication.

Conversely, there are, in fact, several ways in which activists trigger the development of new technologies (Weisskircher, 2019). Animal activists tend to push for changes to the way cosmetics or drugs are developed using animal trials, which would mean significant technology change. Similarly, environmental activists often have specific misgivings about technologies such as windmills or nuclear power plants because they believe they have negative long-term effects on the planet's ecosystem or its purity. Even my own work was briefly censored on the grounds of fear of health misinformation, arguably because I was not an official medical source, until I complained and mysteriously got the ban lifted so I could publish my work on the pandemic.

The impact of social dynamics on technology: case studies

Consider three examples (Segway, the cell phone, Google Glass) of how social dynamics have impacted the success or failure of major technology products.

Technology product failures

Technology is generally viewed as the single force that disrupts markets. However, history is rife with stories of technologies that have failed to meet such hyped expectations. Amazon's Fire phone. Google Glass. Facebook Home. Quikster. New technologies alone don't always cause industry changes. The only technologies that prevail are those that solve a real problem, satisfy a need (or create a "need"), pass regulatory scrutiny, and ultimately find a business model to sustain their roll-out.

Plenty of highly proficient technologies fail not because they don't work well but because not enough people care that they do. Technologies also don't exist in a vacuum. They depend on each other. Most key technologies of the last century have been platform technologies, that is they enable new activity to occur on the network created by the technology being used. Electricity. The internet. Blockchain. 5G. All of these are platform technologies, which means they are systems built upon an architecture that distributes the system into different levels of abstraction and which become used as a base upon which other applications, processes, or technologies are built.

When society isn't ready: the Segway

At the turn of the millennium, an era fueled by technological optimism, serial founder and millionaire Dean Kamen had already developed a myriad of technology consumer products, including the insulin pump AutoSyringe. By the time 1999 came around, he had founded a company based on spinoff technology from an earlier wheelchair product called iBot (Wilson, 2020), having worked up the hype by giving it the codename Ginger (Kemper, 2005). Bedford, New Hampshire-based Segway, producer of a two-wheeled, gyroscopically stabilized, battery-powered personal transportation device, launched its first product, Segway HT ("human transportation"), later marketed as Segway PT ("personal transportation"), on December 3, 2001 as the "world's first self-balancing human transporter," with a speed of 12 miles an hour, which is four times walking speed, turning a half-hour walk into a 7–8 minute ride, essentially attempting to launch an entirely new market. The use case was crystal clear: bridge the "last mile", meaning the way you get from your home to your best public transport option and then from public transport to your workplace. Its technology was impressive. The gyroscopic balancing was peerless, at least when it worked, and user error could be avoided.

Hype and memes you couldn't make up even if you tried

Arguably, the hype around it was similar to the hype around Tesla, the way only millionaires and billionaires can market their new products, through connections to other famous people, in this case through Apple co-founder Steve Jobs, venture capitalist John Doerr, and others, with the message that it was going to "revolutionize transportation forever."

Throughout the years, the company has had three more owners. The year the iconic Hollywood movie *Mall Cop* came out featuring actor Kevin James on his Segway (2009), the founder sold the company to British millionaire entrepreneur James Heselden. In 2010, a year later, the owner of Segway died on a Segway found in a river near his Yorkshire estate, having driven his Segway off a cliff, which didn't help the marketing. In fact, it was a PR nightmare. Medical professionals started noting specific Segway-related injuries. A *National Law Review* article describing California regulations on the matter specifically cited: injuries to the spinal cord, injuries to the head, facial and body lacerations, broken legs and arms, injuries to the head, and permanent disfigurement. Not exactly useful for marketing a product.

After that, the main users continued to be mall cops and tourists, severely limiting the Segway market. In 2013, the next owner, the 45-year-old Brentwood entrepreneur Roger Brown, acquired Segway from a family trust, following the death of Heselden, who had also been a Segway distributor (Capps, 2015). In 2015, its acquisition by Chinese scooter maker and competitor Ninebot was financed by US venture capital group Sequoia and Chinese electronics player Xiaomi.

Fast-forward 19 years, over the summer of 2020, Segway PT, which is the classic version with the handle, having sold only 140,000 units since its launch in 2000, ceased production, and the 21 employees in the Bedford, MA, U.S. plant were let go, although the miniPRO seems to continue. According to Associated Press (2020), the product accounted for less than 1.5 percent of the company's revenue in 2019. Over the past few years, the company had revamped its offering with radically cheaper consumer products, such as the Segway miniPRO Smart Self-Balancing Personal Transporter launched in 2016 with a price tag of $300+, of which my 11-year-old son is a happy owner. Interestingly, removing the handrails makes the Segway infinitely cooler and changes its demographic. No longer is the theoretical limit 100 pounds, now you actually have to have a sense of balance to even try it. That's beginning to sound like aspirational marketing. However, its limitations in form factor still apply: it is bulky, annoying to charge, and cannot just be left on the sidewalk.

Founder Dean Kamen says he believed the Segway would do for walking what the calculator did for pad and pencil. "You'll go further. Anywhere people walk" (CNN, 2018). It was the founder's solution to "the last mile," "getting from the train station to your front door, from your bus stop to your desk at work." The reasons it failed are easy to discuss but hard to determine. Some say Segway tried to be a general-purpose product and failed because it did not focus on any one application (Hartung, 2015). This is an unfair characterization, because the general-purpose aspect persisted, it's just that the only groups testing it out seemed to be a variety of specialist professional groups (city police, mall cops, tourist companies, etc.). Instead, Segway HT/PT's failings are more obvious. Take a look at the product. It's a fat, two-wheel piece of machinery that doesn't stack neatly anywhere. Could you imagine sidewalks filled with them? Even riding hundreds of Segways along busy commuter streets wouldn't work the same way that driving an electric scooter does. In my mind, the form factor alone is to blame for the lack of consumer love.

Hefty price tag of near $5,000 in 2007, but many say it epitomized being a geek, that it just wasn't cool. Plenty of home videos exist on the internet of users falling off their Segways in spectacular yet predictable ways, which tend to be funny to watch and easy to ridicule.

The form factor was not the only issue, Segway also didn't have the business model right. First, it was way too expensive for what it delivered. Paying $5,000 puts it in the category of a used car or is equivalent to something like 200 short taxicab rides or ride shares. Electric scooters have been far more popular for personal mobility, especially in recent years with the rise of the business model of scooter sharing led by Lime and Bird, which has rapidly descended on cities like San Francisco, Tel Aviv, and Paris.

Moreover, the fact that Segway PT might have been overhyped, due to a myriad of issues—the marketability of the founder, the timing of the launch right smack in the middle of the dotcom bubble—doesn't make up for a bad product. That is instructive for people in marketing, because it should serve as a warning to not create hype around a product before you know it has a chance to delight consumers.

Tackling Segway's regulatory constraints

Segway representatives have successfully lobbied states and cities across the United States to allow their use. First, they kept the machine from being classified by the Federal Motor Safety Carrier Administration (FMSCA) either as a motor vehicle or as a scooter. Then, they narrowly avoided the Occupational Safety and Health Administration classifying Segways as powered industrial trucks. Also, the device can be used on private property. The vast majority of states have enacted permissive use of the device. In many US states, Segways are permitted on sidewalks, bike paths, and roads where speeds are slower. Colorado, Connecticut, Massachusetts, North Dakota, and Wyoming have passed laws that do not permit powered conveyances to be operated on sidewalks or bike paths. Typically, the users have to follow the rules of pedestrians when crossing sidewalks and the like. However, due to the limited number of devices in circulation, the case law is limited.

Excellent technology—flawed product

To summarize, Segway HT/PT was a flawed consumer product built on excellent technology. It may even have been overengineered, and certainly seemed to have contained redundant parts that made it almost indestructible, which is an anathema to today's consumer products which

Figure 5.2 Segway's product success vectors

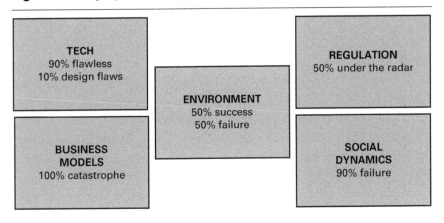

typically have built-in planned obsolescence. To apply our forces of disruption framework (see Figure 5.2), it had technological promise, a horrible business model, an unresolved relationship with regulatory bodies (it was generally neither forbidden nor encouraged—but would have become problematic to local city officials had the sales started to explode), and was a complete disaster from the perspective of social dynamics, because its use case was unclear and the learning curve far from instant. As a general rule, you can fail on one of those five aspects and still fix your business before you are disrupted, but you cannot fail on several at once.

The ways in which Segway did not fail after all

However, before you conclude that everything about Segway was a failure, consider this. The company brand Segway is, perhaps mysteriously, still in use. Segway still owns roughly 400 patents related to personal transporters, covering a wide range of balance-oriented personal transporters, and according to *Wired* even up to 1,000 active patents on self-balancing technologies alone (Wilson, 2020). The founder, Dean Kamen, is also listed with over 1,000 patents, according to Wikipedia. Ninebot, the Chinese robotics startup, uses the Segway brand for a host of transportation devices, from kick scooters via electric motorcycles, an electric hover chair, an egg-shaped seated mobility product unveiled at CES.

When society is ready: the cell phone

Having looked closely at a failure, it would make sense to consider what success look like. There are many technologies to pick from. According to *Wired* (Goode and Calore, 2019), the 10 tech products that defined this decade were WhatsApp, Apple iPad, Uber and Lyft, Instagram, Apple iPhone S, Tesla Model S, Oculus Rift, Amazon Echo, Google Pixel, and SpaceX Falcon Heavy. I own an iPad, iPhone, and multiple Echos; I have driven the Tesla enough to agree; I've taken enough ride shares with Uber to appreciate the value; and I just got an Instagram account to explore. I also have a WhatsApp account but I'm not an avid user.

The most successful technology product in modern times is the cell phone. There are 5.16 billion unique mobile phone users in the world today (GSMA Intelligence, 2020). The number of mobile devices is forecast to grow to 16.8 billion in 2023. I can list a whole host of them as iconic products, beyond the iPhone. Why is the cell phone so successful? Because it maximizes and extends existing social dynamics. Let's explore.

Cell phones were not always that useful. In the beginning, when few had a phone and it was expensive to make a call, the handsets were heavy, and there were no apps, no SMS, and no worldwide penetration and roaming fee treaties, usage was limited. Yet, millions of people had a latent need to make and receive phone calls in many locations where no landline service was available (Carey and Elton, 2010).

Today, however, the use benefits (voice, text, internet) are many, use cases and benefits near instant. Mobile phones are typically easy to use, with little or no learning curve. They require no infrastructure setup by the user because the networks have done that job already. Mobile phones are also status symbols. They are cool. They have their own power supply (battery), which does not rely on the grid unless it is to charge them up. The investments have largely been carried out by private sector providers, which explains why they have occurred rapidly and without delays even throughout many parts of rural Africa. Some 50 percent of Africans have mobile phones, although poor people tend to have very basic versions that do not resemble smartphones, and fixed lines are scarce, making mobile telephony the largest tech platform in Africa. In Africa, cell phones even have an important role in health communication.

Lessons learned from success and failure

Many are those who have speculated on the future of cell phones. What is certain is that the form factor will continue to improve, as will the thousands of technologies and standards that go into it, from screens to processors, to cameras to sensors, to processing units, to the networks on which they operate. Holographic displays have long been popular in sci-fi movies, from *Star Wars* (1977) to *Iron Man* (2008). Bendable screens are now possible through the organic light-emitting diode (OLED) technology. Further personalization will undoubtedly make the cell phone easier to consider as a true extension of ourselves, although the rapid decline of each physical unit and its subsequent replacement ensures that the attachment would have to be to other aspects than the physicality of each device.

Will cell phones continue to be successful or will other devices or technologies take their place? Will personal transportation finally succeed with Segway's successor, the electric scooter? Or will entirely different concepts dominate in the next few decades? Time will tell, but nothing is certain, beyond the fact that ignoring the forces of disruption is a surefire way to have no idea whether you will succeed or fail.

Will cell phones eventually be overtaken by AR devices?

What could unseat the cell phone would be another, more advanced form of communication that provides vastly superior functionality. It might be difficult to match the penetration rates, though, so that technology would truly have to be vastly superior and even then it would take a while for the shift to happen. Could augmented reality be such a product? Not for the foreseeable future and not for the common person. Having said that, the cell phone is "growing old" as a technology. What is more likely is the integration of cell phones with AR, which is starting to happen. AR hardware has typically been expensive and clunky. In marketing lingo, AR is compatible with cell phones, and virtual reality (VR) requires its own hardware; in reality, it is more complicated.

The next step could also involve cell phone implants or some arrangement by which the sensors on the cell phone are further integrated with the human body. Either way, this would not only make the cell phone become a true cyborg extension of ourselves, it would also enable new social opportunities of aspects like eternal connection to another human being on a sensory level at a distance and such interesting prospects. Whatever solution is chosen, it clear that the issues of tracking and social acceptance will become even more important to discuss and will decide the ultimate mass market success of such "enhancements." Ultimately, the cell phone's successor may become more of a cognitive all-in device capable of suggesting or carrying out your responses ahead of time rather than merely a real-time communication layer. That's when the entire notion of what it means to be a social being will come into question and we will start to have a whole other debate about AI-enhanced social dynamics, which may gradually morph into AI-driven dynamics, which would blur the boundary between natural and synthetic intentionality. That is a few years ahead, but it belongs to the kind of issues we need to consider in a chapter on social dynamics.

The question I ask in this chapter is not so much what technology will replace another but how social dynamics interject with any new technology's path in powerful ways. What cell phones have going for them is that they enable mass-scale, personalized communication. What the technology ended up doing, which was not at all determined at the outset, was to almost literally swallow up another pathbreaking technology, that of the internet, in its march toward domination. Whatever technology comes along, that is a tall order, and one not accomplished by niche tech prowess. For that reason alone, cell phones with their enhancements, and not the fixed, tethered internet, are the lasting mega-technology of the next millennium. The technology that would want to unseat it would have to provide a more direct yet mass-scale, ubiquitous way to reduce the latency and increase the communication bandwidth between people. As of now, there's only one candidate, and it almost failed, and that is another wearable form factor, namely smart glasses.

When society isn't ready, but enterprise is: Google Glass

Considering the evolving story of Google Glass is instructive in terms of understanding the social dynamics of technology. Glass was the brainchild of Georgia Tech's Thad Starner, a pioneer in wearable computing. It was conceived as an augmented reality overlaying the wearers' field of vision with real-time information about their surroundings. Out of the gate, the functions included an internet browser, camera, maps, calendar, and other apps by voice commands.

The initial challenge with this promise was to enable such functionality seamlessly. Processing visual information is still data intensive and may be a few Moore's Law generations away in terms of battery capacity, image recognition, and telecommunications Gs in terms of network speed. In other words, it wasn't the use case that was the problem (although that was also a problem because they did not properly choose one) but the timing of the launch and the brand promise.

After being launched with much bravado in 2013 as experience augmentation—after all, the 2012 demo reel featured skydiving, biking, and wall scaling—it failed miserably as a consumer product quickly after launch and was pulled from the market in 2015, for well-understood reasons. Statista estimated that more than 831,000 units were sold in 2014. It was targeted toward a general public audience but had a high price tag of $1,500, which gave it the impression of being a super-premium product when in fact it was more like a prototype. The main functions were to check messages, view photos, and search the internet, functions for which the products it competed with, notably cell phones, did a much better job at very little annoyance to most users.

Can you succeed with poor design form factor?

Glass had clunky design—perhaps not an aesthetic design that might turn most people off, as critics claimed, but not one that would attract them to wearing them as an outright fashion statement, as is the case with some designer glasses. There were concerns about privacy, given that the glasses would record everyday interactions (this was never agreed with authorities) or systematically recognize strangers (for personal or professional use by law enforcement), as well as about safety when operating personal mobility devices (from cars to scooters).

The "release often, release early" mantra of software does not hold with hardware consumer products, which is something Sergey Brin, Google's founder, did not fully take on board. Also, Glass was not truly augmented reality, it was more like supplemented reality—it was definitely more vitamin than cure, to use the metaphor we earlier dissected. In fact, it did not do any individual action especially well. With its limited functionality, it interfered more than it helped regular social interaction, at least when not everybody was wearing the glasses. In reality, smart glasses initially failed in the market due to lack of clarity on why the product existed.

Pivoting from consumer to enterprise

Today, Google Glass is an enterprise product. I was aware of this only because I know entrepreneur Ned Sahin from MIT at his startup Brain Power. Google Glass Enterprise Edition has already been successfully used by Dr. Sahin to help children with autism learn social skills, which proves the point that the technology is more than sufficient for specific medical and professional use cases. Google Glass Enterprise Edition 2 was launched in 2019 with a new design, new specs, and a $999 price tag. The enterprise use cases are highly specific: doctors, warehouse managers, maintenance workers who when fixing machinery are able to focus on the job in front of them while simultaneously having access to a manual, factory workers, farm workers, and more. For instance, it might allow physicians to live-stream the patient visit; it could allow surgical students to more accurately see what their teachers are seeing as they operate; it could offer real-time lactation consulting for breastfeeding mothers; or other family support services; each a slightly unsexy use case compared with the fashion icon-clad launch in 2013. Broadly, the device would arguably reduce mistakes, increase efficiency, and improve task focus and concentration on detail-oriented tasks where previously one had to stop the activity to take notes, record the activity, or look something up. Forrester predicts something like 14 million workers could be wearing Glass in this capacity.

In each case, what matters to the wearer is not interaction with other people, but interaction with other technology and remote contexts where social interaction is a rarity, not the norm. In such conditions, Glass augments the user's ability to do their job and does so without hampering their social interactions. The emphasis is on safety and efficiency rather on communication, which is very different.

Glass is the only true outcome of Google X, although the R&D company now simply called X could still surprise us. No matter what you might think about its failure or not, Google elevated public awareness of smart glasses to unprecedented levels. Smart wearables are definitely here to stay, even if it takes another decade to match the form factor with the intuitive, impressive functionality that outcompetes or complements the cell phone and other mobile technologies.

Could Google or others return to the consumer smart glasses market any time soon? Rumors always exist. A number of vendors would have potential game-changing interfaces or technology to bring to bear, notably Sony, Google, Microsoft, Epson, Toshiba, Qualcomm, Recon, Vuzix, APX, and CastAR, among others. Market estimates and projections for the smart glass market over the next 5–10 years vary wildly between $5 billion and $100 billion, to the point where it's pretty obvious that nobody in the market research industry has any inkling as to what might be going on in that market. That is, in part, what make smart glasses fascinating and important to consider as a future technology. However, I think it's important to realize that the fate of augmented reality does not stand or fall with smart glasses as such. AR will likely require its own product interface but will also be integrated with most other form factors in some way.

Regardless who presents it and when the smart glasses concept returns, a killer application functionality and must-have would have to be some kind of awesome augmented reality that does something the cell phone cannot easily do and which, at the same time, does not feel creepy. Tall order? Perhaps, but this is what a product designer has to contend with in order to succeed.

Emerging interfaces to the internet

Finally, there are several other technologies that are contenders for a new interface to the internet, notably AlterEgo, MIT Media Lab student Arnav Kapur's system to surf the internet with his mind. AlterEgo (2020) enables the users to silently Google questions and hear the answers through vibrations transmitted through their skull and into their inner ear. However, that device, as useful as it might become for people with speech disorders, is not currently envisaged as a general-purpose communication device, although that is clearly a potential application.

Conclusion

Misunderstanding or ignoring social dynamics is a major reason why advanced technology initially (or sometimes definitely) fails to capture its intended users' imagination and as a result does not integrate into their lives. In this chapter, I looked at Segway, Google Glass, and cell phones. Plenty of other examples would have been equally enlightening in this regard and future and emerging examples unfortunately will persist. Going forward, I would expect companies, product developers, and governments to redouble their efforts toward understanding social determinants of use.

Having said that, it is also the case that if innovation was purely left to intended consumers, we might not experience the radical innovations that characterize founders who defy common wisdom, focus groups, and expert advice and simply develop something according to their own, deeper vision. Those products are rare, and they rarely succeed, but when they do, such as a few of Steve Jobs' Apple products, they succeed in surprising, society-altering ways.

For sure, no matter how much we try to imagine or test what real-life human beings do when confronted with a technology product, we cannot predict it fully.

Key takeaways and reflections

1 Reflect on which main social group(s) you belong to and the potential impact those groups have on technology. To what extent have you participated in group activity to engage or voice your opinion on a technology product you own or have an opinion about?

2 Pick a technology you care about in terms of its adoption cycle over the next decade. Map its likely trajectory in the four generations that dominate consumption (Boomers, Millennials, Gen X, and Gen Z). At what point do you envision this technology will be replaced by another?

3 Imagine creating a social movement around an embryonic technology— either as a reaction to it or as a support group favoring its further adoption. Who would you seek support from? How would you build the movement? What would your goals be?

4 Imagine we are in 2030, looking back at the past decade. What were the main social movements that characterized the decade? How did they emerge? What technologies did they make use of?

Five technologies that matter and why: AI, blockchain, robotics, synthetic biology, and 3D printing

In this chapter, I dive into the five technologies that matter the most at this moment in time: artificial intelligence, blockchain, robotics, synthetic biology, and 3D printing. I describe how each technology is embedded with a complex set of forces of disruption (sci-tech, business models, policy and regulation, social dynamics, and the environment), what each technology is capable of today, and what it might achieve in 10 years. In the appendix, you can find shortlists of (a) who you should track (scientists, innovators, startup founders), (b) what you should read (publications), and (c) which tech conferences you should attend (virtual or in person), to have any inkling about where it truly is evolving day by day, year by year for each of these emerging technologies.

Tech changes may seem bewildering. In my experience, most executives can manage only a small number of priorities at any given time. I believe there are five key technologies that matter at this time: artificial intelligence, blockchain, robotics, synthetic biology, and 3D printing.

I chose these from among other emerging technologies because these technologies interact with each other, creating previously unthinkable conditions of technological, biological, material, social, and psychological change. Using these technologies, items as commonplace and well known as a wall, a piece of cloth, and a human being may become nearly unrecognizable from their predecessors over the past 1,000 years within the span of this decade alone.

Each of these five technologies also has a unique contribution to the post-pandemic world. AI is enabling analytics on epidemiologic patterns, drug discovery, advanced tracking of the supply chain, and more. Blockchain is empowering the next generation of peer-to-peer economic exchange in a myriad of fields, allowing this activity to occur without a middleman, thus encouraging the kind of distributed activity we need to foster in a post-centralized model where humans need to spread out to avoid contagion. Synthetic biology is enabling a much more efficient way to produce vaccines in the future and potentially could speed up clinical trials. 3D printing is enabling distributed production and consumption of manufactured items without going through a vulnerable, physical supply chain.

The success of these technologies in helping address the pandemic will depend on how well we support the scientists who are attempting these feats. In order to facilitate the process, regulators will need to have a fundamental grasp of the impending changes in each core technology and would need to develop a thesis on how they think the technologies would evolve and interact. Politicians will need to provide programmatic R&D support across a wide array of sci-tech domains, not all of which will see immediate results. Businesses will have to pivot both to react to and to make the most of such opportunities and challenges. Consumers may have to endure periods of drought as these technologies mature. In short, there are no guarantees with technologies of the future. Their success will ultimately depend on how well they are able to embed in our society, and how they latch on to the existing patchwork of solutions that we already use.

The future of artificial intelligence

Artificial intelligence, as I define it, is the ability of a computer program or a machine to learn and solve problems. Simply put, I define artificial intelligence as machine learning. While we are far away from the science fiction-based world in which computers take over society, it is important to note recent events in which machines have been able to outthink humans. For example, in 2016, AlphaGo by DeepMind defeated the human world Go champion Lee Sedol. This is a notable accomplishment. An ancient Chinese board game, Go is known as the most challenging classical game for artificial intelligence because of its complexity, with exactly 10 to the power of 170 possible board configurations, which is more than the number of atoms in the known universe. Go is based on stringent rules and fits the types of challenges that we can reasonably expect computers to be willing to engage with today. Over recent years, such systems have started to overtake humans for all classical games, notably chess, Go, and Shogi, extending to the early computer games from Atari and even some contemporary ones such as Dota 2.

The future success of AI is dependent on the impact of the four forces, namely the science and technology community, its application in business, the regulatory environment, and how humans (society) adopt it.

The impact of the scientific community on AI

Given the depth of knowledge required, advances in the field of AI are dependent on the work of scientists with PhDs from leading universities. The possible exception is larger multinationals which may, in their various divisions, be exploring AI to accomplish business objectives but where the C-level leadership is of a previous generation of tech literacy, which may mean that they can only scratch the surface of what the technology really is capable of.

The impact of business models on AI

AI becomes productive in a business context when it can be used to solve problems that consumers will pay for. Earlier generations of AI were abandoned because of the lack of results beyond initial domain-specific breakthroughs (beating humans at chess) but also because of an ability to scale in a useful way without too much tinkering. Interpretability is also key to a business application, so unless the approach taken by the machine can be understood and explained to laypersons at least in some simplified way, it will not be trusted to carry out business.

The impact of regulation on AI

Because AI touches on such sensitive topics as privacy and security—think robotics and autonomous driving—it has drawn the attention of policy makers. As West and Allen (2020) write in *Turning Point: Policymaking in the Era of Artificial Intelligence*, near-term policy decisions could determine whether the technology leads to utopia or dystopia. They recommend creating ethical principles, strengthening government oversight, defining corporate culpability, establishing advisory boards at federal agencies, using third-party audits to reduce biases inherent in algorithms, tightening personal privacy requirements, using insurance to mitigate exposure to AI risks, broadening decision making about AI uses and procedures, penalizing malicious uses of new technologies, and taking proactive steps to address how artificial intelligence affects the workforce. All of that and more will be needed. Most of all, we need to reskill our politicians. Without that, any regulatory activity is likely to backfire at some point.

As AI becomes embedded in more and more industrial applications, the scrutiny will only increase. In the United States, the US Committee on Foreign Investment in the United States (CFIUS) has classified AI as a sensitive technology for national security purposes. In 2018, Congress enacted the Foreign Investment Risk Review Modernization Act (FIRRMA), expanding its jurisdiction to cover so-called non-passive, non-controlling investments in US companies in three focal areas: critical technologies, critical infrastructure, and sensitive personal data (CRS, 2020). In January 2020, the US government imposed certain limits on exports of artificial intelligence software, according to Reuters (2020). The measure covers software that could be used by sensors, drones, and satellites to automate the process of identifying targets for both military and civilian ends, but stops short of a more general export prohibition which would have jeopardized wider markets for exports of AI hardware and software. The CFIUS regulation also covers a mandate for the submission of a declaration to CFIUS for transactions involving certain investments related to critical technologies (Dohale et al., 2020). The final CFIUS rules include "emerging and foundational technologies" (which includes AI) in the definition of "critical technologies (Carnegie et al., 2020).

In Europe, the General Data Protection Regulation (GDPR) was issued in 2018 to regulate AI in the handling of personal data. Now, the European Commission (EC) is addressing the issue of industrial data. In February 2020, the EC released a White Paper that seeks to ensure societal safeguards for "high-risk" artificial intelligence (EC, 2020a). The commission emphasizes how AI models affect millions of people through critical decisions related to credit approval, insurance claims, health interventions, pretrial release, hiring, firing, and much more (Engler, 2020). According to the EU, there is the potential for artificial intelligence to "lead to breaches of fundamental rights," such as bias, suppression of dissent, and lack of privacy. The report calls out sectors such as transportation, healthcare, energy, employment, and remote biometric identification.

One area that the commission is particularly concerned about is facial recognition. A leaked draft white Paper suggested a moratorium on facial recognition in public spaces for five years, but the final paper calls only for a "broad European debate" on facial recognition policy (Chen, 2020). The EU argues that responsible regulation will build public trust in AI, allowing companies to build automated systems without losing the confidence of their customers. It suggests legal requirements such as making sure AI is trained on representative data, requiring companies to keep detailed documentation of how the AI was developed, telling citizens when they are interacting with an AI, and requiring human oversight for AI systems (Chen, 2020). In the White Paper, the EU signals its intent to increase its investment to match that of the US and China, along with an emphasis on skills as well as small and medium-size enterprises.

What does the Chinese regulatory regime for AI look like? Back in 2017, China's State Council ambitiously declared its objective to become the world leader in AI by 2030, to monetize AI into a trillion-yuan ($150 billion) industry, and to emerge as the driving force in defining ethical norms and standards for AI (Roberts et al., 2019). China has set up five national AI platforms (automated driving, intelligent city, medical imaging, intelligent audio, and intelligent vision), each with a large domestic tech company receiving preferential contract bidding, easier access to finance, and (potentially) market share protection and has set as its target coming up with a comprehensive AI regulation by 2025 (Ning and Wu, 2020). Finally, it is quite clear that China sees AI as its opportunity to leapfrog and potentially eclipse the US in military technology and, perhaps ultimately, power, through asymmetric capabilities (Roberts et al., 2019).

Figure 6.1 AI as a force of disruption

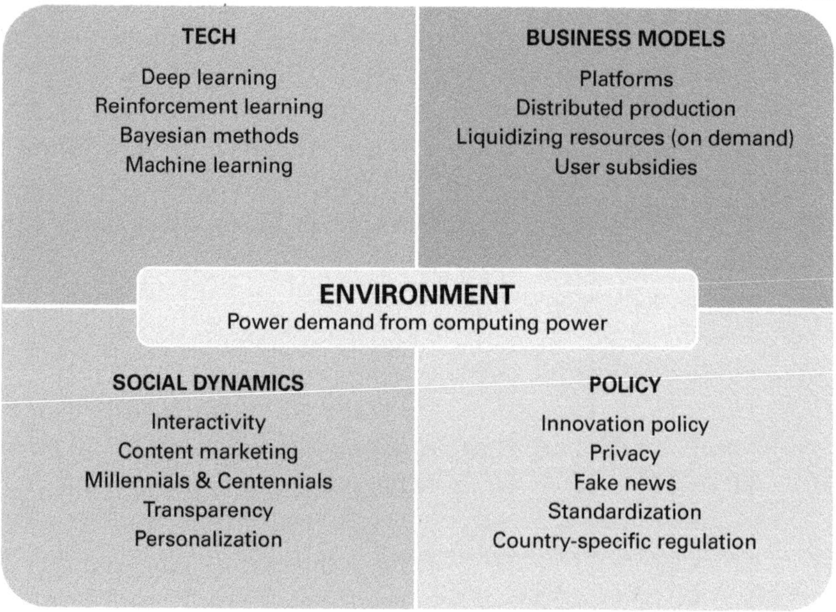

TECH	BUSINESS MODELS
Deep learning	Platforms
Reinforcement learning	Distributed production
Bayesian methods	Liquidizing resources (on demand)
Machine learning	User subsidies

ENVIRONMENT
Power demand from computing power

SOCIAL DYNAMICS	POLICY
Interactivity	Innovation policy
Content marketing	Privacy
Millennials & Centennials	Fake news
Transparency	Standardization
Personalization	Country-specific regulation

The impact of social dynamics on AI

AI is deeply intertwined with social dynamics in several ways (see Figure 6.1). To have real-world accuracy and relevance, AI models are often trained on data sets that were prepared by people. This means an army of people is actively involved in labeling data sets, whether it be labeling images, classifying terms, products, or otherwise. The principles with which they do that will determine how productive (or even misleading sometimes) the resulting AI models will become.

What AI technology is capable of today

Today, AI is capable of fetching and structuring information from a huge body of evidence the way no human could and certainly at higher speeds, but so far in limited domains only. If you have a trained AI model with 100,000 images it could likely find all cat images faster than a human would, but mostly because it was trained on the classifications done by dozens of humans before (although some novel AI systems learn important distinctions without human mentoring).

Similarly, if you have a bunch of unstructured information but were looking for something quite specific, with a trained AI model you might be able to get results that few humans could spot. This approach was recently tried to discover potential chemical compounds that might be relevant for a potential coronavirus vaccine.

AI is also truly useful in optimization tasks such as weighting different options for how to route goods in a complicated supply chain, how to organize traffic, etc. Eventually, AI will be great at autonomous driving, particularly as smart road infrastructure improves signage. Currently, the experimental systems out there are sufficient for limited tasks such as lane assist, but anything else is hazardous (if legal).

AI is beginning to yield its force in spheres previously unthinkable, including the evolving perceptive AI experiments that are pioneered by MIT startup Affectiva, which started with analyzing emotions of people watching ads based on manually labeling huge data sets with facial expressions and adding machine learning analysis on top, with an emphasis on automotive use cases such as environmental perception of the driver, passengers, and the physical environment inside the car. Its perceptive AI might be deployed for all kinds of creative scenarios in the near future, perhaps the next 1–3 years. Anywhere that needs AI-enhanced perception could be up for grabs, whether it is a speaker wanting feedback from online conference audiences, online focus groups for product feedback, or any other remote scenario where real-time analysis of social cues is required.

What AI might achieve in 10 years

Looking at AI in a decade's horizon, the usefulness will drastically increase but likely not exponentially. We might be able to use it for near full autonomous driving on highways and in city centers, at least if we invest in the hardware (Tesla's and that of its emerging competitors) and intelligent road infrastructure, which will in part depend on government investment and regulatory approvals. We will be able to rely on it for most computing tasks. AI will be embedded in decision software in most industries but may not yet be pervasive. The challenges with speech recognition will still be there but will be resolved for most common applications. AI will be a formidable competitor (or assistant) in all structured games or indeed in any rule-based scenario.

There are two ways of thinking about how to track AI. The first is to study the evolution of the basic technology by tracking computer scientists who are active in the field, educating yourself with readings and attending conferences. I provide recommendations on these in the appendix. The other approach is to study the main application areas, notably the ones you regularly follow because of your own business interests as well as a few emerging ones that are important to understand the potential. The challenge is to pick which those areas will be and to switch them out when others become more prominent. A starting point might be to browse market research and consulting accounts of "top trending" technologies in their annual publications, but that is truly just a starting point. These documents are rarely worth more than 10 minutes of your time. Instead, what matters is what is being said and done by domain experts in the technology or in the application area you focus on.

The future of blockchain

Blockchain is a nascent financial technology in existence for less than 10 years which has the potential to become the new platform technology for financial transactions. Think of it as a type of digital bookkeeping, except that the records going into it cannot be edited or deleted and the data can be kept anonymized in numerical format. Blockchain provides a decentralized peer-to-peer network to maintain a record of transfers from one party to another party. The integrity of those transactions, and the ledger itself, are maintained by cryptographic tools. Additionally, a blockchain entails a protocol with consensus mechanisms that verify the data being stored or transmitted and contains rules and procedures for determining if, when, and how to update the ledger (Dewey, 2019:6).

Fundamentally, blockchain is about a new financial paradigm built on a peer-to-peer model of no intermediaries. The main benefits are trust and efficiency. By enhancing trust, blockchain may make the system it enables more efficient. Conversely, efficient systems are typically trusted more because they deliver high value, so the two benefits are mutually reinforcing. After initial use cases in finance, blockchain is being tested out in nearly every industry, from financial services and payments to healthcare, energy, and property (even intellectual property) management.

All forces of disruption interplay in deciding what blockchain will become. Both the speed and direction are wholly unclear, but a few things seem certain: it will happen no matter if there is resistance from incumbents, it will be disruptive for everyone involved, and it will challenge existing power structures. Having said that, the rest will be determined by the exact interplay of the forces of disruption I will now describe.

The impact of regulation on blockchain

The main obstacle to the success of blockchain is regulation. Today, regulators struggle with the consequences of admitting blockchain to the heart of finance, and this uncertainty creates a destabilizing undercurrent which can undercut further innovation of this technology. Unfortunately, the default, knee-jerk reaction of many regulators is not to engage with and understand the potential of blockchain but rather to worry about its potential criminal application.

A handful of countries, including China, Russia, and Colombia, have banned Bitcoin and other cryptocurrencies (Dolan, 2020). The US has taken a more permissive stance. A full 32 states have blockchain-related legislation (Morton, 2019). However, in June 2009, Facebook, of all companies, proposed a blockchain cryptocurrency called Libra, to be accompanied by a digital wallet called Calibra. From that day, it became entrenched in debates with lawmakers, notably the US Senate. This caused companies like PayPal, eBay, Mastercard, Strip, and Visa to withdraw from the Libra Foundation even before launch.

The EU's Fifth Anti-Money Laundering Directive (5AMLD) was signed into law in January 2020, marking the first time that cryptocurrencies and crypto services providers fell under regulatory scrutiny.

The impact of business models on blockchain

As promising as the technology use case might be, there are significant challenges to surmount before blockchain goes mainstream. Blockchain vendors still do not speak the language of mainstream business. They are stuck in tech terminology where the market demands user value and efficiency. Data written on blockchain requires standardization for the system as a whole to work well and to allow for interoperability between blockchains.

Other challenges to the success of blockchain include cost, the issue of patents and IP, and regulatory bodies needing to be onboarded. Even though there is promise, not only do security issues need to be resolved technically, the trust needs to permeate the economic system. These things will take time (Yoshi, 2020).

The impact of social dynamics on blockchain

Blockchain will not succeed unless a great many consumers, voters, and stakeholder groups understand it, see its potential, and start experimenting with it. Considering how complicated blockchain is to explain and demonstrate today, this is a considerable challenge.

The next decade will be spent proving the use of blockchain in various scenarios (see Figure 6.2). In the process, one will find out whether a great tech concept can survive the scrutiny of social groups that have objected to numerous technologies before.

To a huge extent, a people's first exposure to blockchain will color their response. If you find out it had something to do with blocking your loan application, that will adversely affect your view of it. If, on the other hand, you find out that it leads to less corruption in government, it might put the technology in a positive light. Blockchain can, of course, do both. It can also be used for nefarious purposes; much like all technology, it is a double-edged sword. Transactions over blockchain can indeed be monitored but they are also encrypted, which means beyond what is being exchanged and the amounts, you cannot see exactly who is exchanging. It provides a layer of

Figure 6.2 Blockchain as a force of disruption

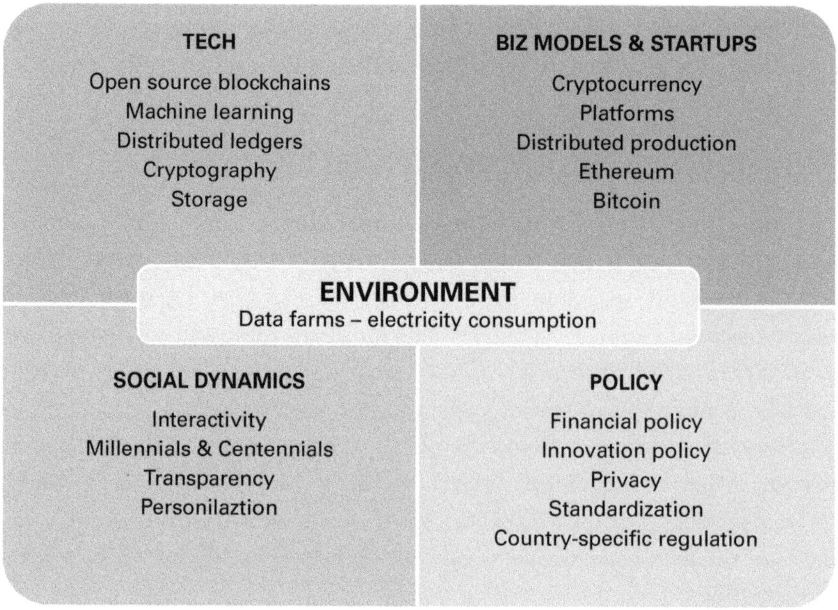

TECH	BIZ MODELS & STARTUPS
Open source blockchains	Cryptocurrency
Machine learning	Platforms
Distributed ledgers	Distributed production
Cryptography	Ethereum
Storage	Bitcoin

ENVIRONMENT
Data farms – electricity consumption

SOCIAL DYNAMICS	POLICY
Interactivity	Financial policy
Millennials & Centennials	Innovation policy
Transparency	Privacy
Personilaztion	Standardization
	Country-specific regulation

financial transactions that potentially could happen right under the eyes of law enforcement and prosecutors.

How the public will react as they understand more about blockchain is an empirical question: We will simply find out. Slow and steady would be the most successful approach if you are in favor; rapid and sweeping would likely lead to a backlash.

What blockchain technology is capable of today

Today, blockchain powers niche fintech applications and is being road-tested for potentially taking on key financial functions over the next decade. Will it replace the traditional financial centralization and gatekeeper function played by banks? Not likely, but it will drastically change the way they operate.

Some governments, such as in Venezuela with Petro, a government-backed digital currency aimed at being an alternative to the country's battered currency, and the Brazilian region of Bahia (Macaulay, 2019) are starting to experiment with how blockchain can help reduce corruption in public bidding. Estonia is the country farthest along, and was also the earliest adopter, given its advanced e-government infrastructure dating back over a decade. Its KSI blockchain system secures most of the information passing between government and citizens. Dubai is in the process of digitizing the entire city on the blockchain, starting with a decentralized learning platform, going paperless, and deploying a shared platform for all government entities, arguably aiming to unlock annual savings of 5.5 billion dirhams ($1.5 billion) based on document processing alone (Consensys, 2020; Smart Dubai, 2020).

What blockchain might achieve in 10 years

Today, blockchain has been used for subscription payments using cryptocurrency instead of fiat, improving the UK land registry, integrating rural banks into the Philippine banking system, and track and trace of luxury goods (Consensys, 2020). Gartner predicts blockchain will create $3.1 trillion in business value by 2030 (Gartner, 2019). Other analyst firms have wildly diverging predictions. In some ways it does not matter. The key thing to keep in mind is that the numbers are large and the market is expanding rapidly, with the likely opportunity and disruption that must logically follow, but exactly how is the question.

Within the next decade, it will arguably become the default fintech approach, it will underpin our financial system, and it will create the backbone of an entirely new set of business models in finance, but also in a plethora of other sectors, perhaps almost all sectors.

The main general use case is verification. Verification is the key to any market: goods, products, services (employment, contractors, procurement), by enhancing security, standards and compliance, interoperability, transparency, providing supply chain monitoring, tracking, and ensuring a fair marketplace.

The HR industry is an area well suited for blockchain, with payroll as a top use case. Using blockchain makes it possible to employ and pay workers in remote locations or countries where payment infrastructure is limited or fiat currencies are volatile (Gartner, 2019). The service is in its infancy. Bitwage (https://www.bitwage.com/) is an example of a startup that, since 2019, helps companies pay their employees in cryptocurrency. Perhaps more importantly, because blockchain is designed for trust (Mercer, 2019), verifying CVs and personnel records as well as assisting in background checks might become the true killer application.

Legal agreements are a paradigmatic blockchain use case. The benefit is clear: a smart contract can never be changed, and no one can tamper with or break a contract. In 2017, Belarus became the first country to legalize smart contracts and a number of US states—Arizona, Nevada, Ohio, and Tennessee—have passed legislation, too. The EU is building a services infrastructure and is actively studying the impact of blockchain, aiming to become proactive (EU, 2020). Currently, four use cases are selected: notarization, diplomas, self-sovereign identity across borders, and trusted data sharing.

Life science can use blockchain to make aspects of clinical trials as well as drugs in market more trusted, ensuring authenticity along the value chain from R&D to medical application. The ability to track every aspect of a life science product is immensely valuable and is not at all implemented yet. Already, the US Food and Drug Administration (FDA) and IBM are collaborating on using blockchain to identify, track, and trace prescription medicines and vaccines distributed throughout the country. Another initiative is the FDA's MediLedger pilot project in partnership with 25 pharmaceutical companies trying to meet the 2023 requirements of the Drug Supply Chain Security Act (DSCSA), which is to require tracking of legal changes of ownership of pharmaceuticals in the supply chain.

A significant amount of the data created during late-stage preclinical and clinical phases of drug development is submitted electronically to regulators prior to drug approval. Companies need to prove to regulators that the data has not been tampered with since it was first obtained. In 2019, Boehringer Ingelheim agreed to manage its clinical trials on blockchain technology provided by IBM Canada. Scientist.com, a marketplace for outsourced research, is now also using the technology to do just that (Lynch, 2019).

In retail, blockchain can reduce the impact and presence of counterfeit goods, particularly in the luxury segment and particularly in fashion. The impact on supply chains would be transformational. In media and entertainment, blockchain can drastically reduce piracy of intellectual property, protect digital content, and facilitate the distribution of authentic digital collectibles. Blockchain has the potential to disrupt social media by giving users an opportunity to own and control their images and content. Currently, when you share content, on most platforms you give up the right to that content. Even if you do not, there is little way to control its re-use. In sports and esports, blockchain can increase fan engagement with enhanced loyalty programs and incentives.

Coverage of a myriad of services and functions of government that previously may have been attempted by e-government approaches are ripe for transformation through the creative use of blockchain. Some examples include taxes, law enforcement (e.g. criminal records), public procurement, voting records, military records, pension records, healthcare records, welfare records, secure data sharing, data personalization, and more. These areas need to constantly verify that their records are correct, need to update their records, need to give access to certain records, not to others, and need to analyze those records to provide services on top. Each is possibly enhanced by the blockchain approach. Imagine there was a way to track vaccinations across the globe, so you could know exactly who had which vaccine and when the vaccine would need renewal. Think about all the effort going into food inspections and the poor compliance we still see, given that you cannot control and monitor each establishment.

Blockchain is poised for significant growth in the next decade, both in overall value and in terms of its importance in society. There are indications the technology will be strongly implemented in and around the world's top financial capitals (London, New York, Singapore, Hong Kong) and also in emerging markets, given the potential to leapfrog building out a traditional financial infrastructure (the way mobile made its way to Africa before landlines did).

The future of robotics

Robots are hardware that sense, "think," act, and communicate on our behalf. Thousands of sensors are available for such systems today. Robots do not really think yet, but they do run algorithms that enable them to do things like recognize objects, understand natural language (to some extent), and, at times, create content. Robots today are highly contextual—in other words, they cannot operate very independently across different contexts unless those contexts are highly configured.

Industrial robotics has increasingly gone mainstream. During the COVID-19 pandemic, many manufacturers went from being curious or testing robotic technology to deploying it. Consumer robotics is a much smaller field; the only well-known player is Massachusetts-based iRobot, with its set of robotic vacuum cleaners. The other types of robots on the market include professional service robots, found outside factory manufacturing environments, used to automate dangerous, time-consuming, or repetitive tasks, freeing humans for cognitive functions; military robots, deployed in combat support for anti-submarine operations, laying mines, fire support, electronic warfare, battle damage management, strike missions, aerial refueling; and security robots, designed to replace security guards, control property, and provide mobile CCTV monitoring.

The impact of regulation on robotics

As robots arguably become safer, the entire regulatory framework around robots needs to be reviewed, especially around issues concerning liability, privacy, and autonomy. Robotic surgery is covered by medical device regulation in most jurisdictions. The European Union's Medical Device Regulation (MDR), requiring all products to obtain the CE mark, will apply in May 2021, perhaps taking half of the market offline and leading to major innovations in order to comply and compete in the new environment (Schweighart, 2020). In the US, regulators are traditionally most worried about safety, and in China regulators worry about foreign technology. The EU tackled the issue early, with an R&D project called RoboLaw (2014), which looked at guidelines in four areas: driverless vehicles, robotic prostheses (and exoskeletons), surgical robots, and robot companions. The project emphasized the importance of agreeing on definitions of what a robot is (and not) and emphasized standardization. In 2019, the European Parliament adopted a comprehensive European industrial policy on artificial intelligence and robotics (EU, 2019). One unresolved issue is whether robots will be granted "personality" and legal status just like organizations.

The impact of business models on robotics

Industrial adoption is now the biggest battleground for robotics and is likely to continue to be so throughout this decade. Given that, the complex set of forces of disruption (sci-tech, business models, policy and regulation, and social dynamics) will entail close interaction between corporations along the value chain as well as government.

The impact of social dynamics on robotics

While humans are attracted to the idea of robots, we fear the consequences of their presence, likely due to the images portrayed in science fiction books and movies.

Social acceptance of robots on the factory floor will be greatly enhanced by the fact that robots increasingly take the dangerous, repetitive, and heavy lifting out of frontline work. There is an increasing number of robots that are co-bots, e.g. they collaborate with humans rather than doing things completely separate from them. When that interaction works well, it builds goodwill, among the executives who have put the co-bots in place and in terms of the "sentiment" between robots and humans.

Over and beyond these observations, the entirety of Chapter 9 is dedicated to the man/machine symbiosis, so I have gone through some further perspectives regarding the risks and rewards in that chapter (see Figure 6.3).

What can robots do today?

The main application of robotics today is industrial robots in a warehouse or factory that either assemble products based on highly standardized moves or pick out merchandise from shelves and bring it to the next part of the logistics chain. Kiva Systems (2003) was a Massachusetts-based company that manufactured mobile robotic fulfillment systems. The whole system is designed to make warehouse stock mobile. The company was bought by Amazon for $775 million in 2012 and is now known as Amazon Robotics. Amazon has 200,000 robots working in its warehouses and as a result of the

Figure 6.3 Robotics as a force of disruption

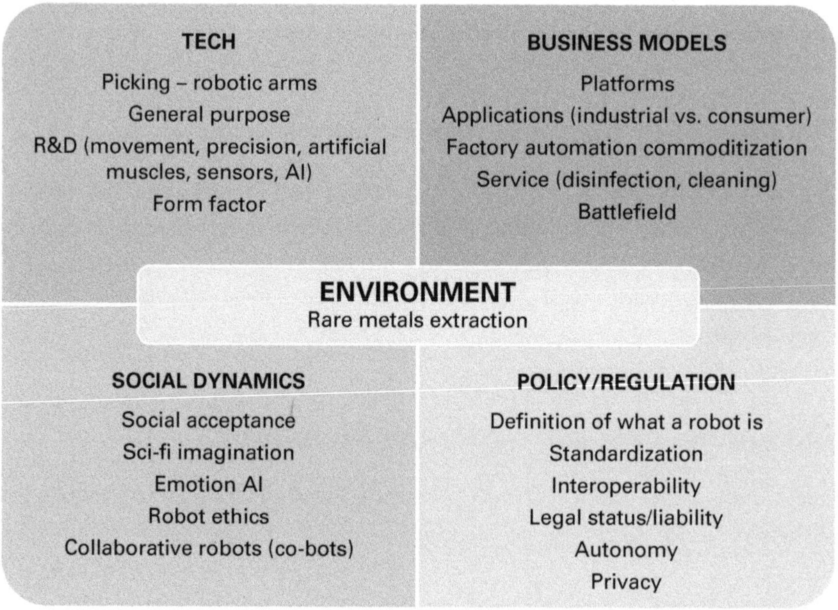

TECH	BUSINESS MODELS
Picking – robotic arms	Platforms
General purpose	Applications (industrial vs. consumer)
R&D (movement, precision, artificial muscles, sensors, AI)	Factory automation commoditization
Form factor	Service (disinfection, cleaning)
	Battlefield

ENVIRONMENT
Rare metals extraction

SOCIAL DYNAMICS	POLICY/REGULATION
Social acceptance	Definition of what a robot is
Sci-fi imagination	Standardization
Emotion AI	Interoperability
Robot ethics	Legal status/liability
Collaborative robots (co-bots)	Autonomy
	Privacy

new robotic capability and surging demand, Morgan Stanley expects Amazon to ship 6.5 billion packages in 2022 (Edwards, 2020).

Robotic process automation (RPA) systems are another recent robotics story, which is more about digital automation but a clever marketeer appropriated the robotics label. RPA allows firms to configure entire sets of processes that previously were manual steps into a chain of automatically triggered events. So far, the use cases have been customer service, payroll processing and sales order preparation, and information storage, sales invoice processing, tech support, as well as personal use, but many others are foreseen (AI Multiple, 2020).

Both factory automation and surgical robots are in controlled environments. They are performing repetitive motions 24/7 but do little thinking.

Robotics is going mainstream. COVID-19 has taken it to application, scale, and not just testing.

What robotics might achieve in 10 years

Over the next decade, robots will be able to do many more things, in both the industrial and consumer realms. In industry, they will increasingly be capable of multistep functions (not just logistics, picking, or single-function motions) as well as monitoring, hygiene, and maintenance functions. Drones will increasingly take on robotic capability and will be able to pick up, fly, and drop off goods under a variety of conditions.

In the consumer realm, the vacuuming, disinfection, and cleaning robots will proliferate, especially in light of COVID-19. Will they take millions of jobs? For sure they will! But how many more jobs will be created in the meantime that still require humans is the real issue. Any job that requires cognitive skills, moving around between many different locations and contexts, or socio-emotional skills such as interaction, care, or training, would also seem to be safe for now, although aspects of those functions can surely be automated. Robots are co-workers more than a replacement (Smit et al., 2020).

It is possible that bio-inspired robots will become more than prototypes throughout the decade. Work is ongoing in the area of artificial muscles, embedding living cells into robots. If we do make major progress in brain–computer interfaces (e.g. Neuralink and other approaches), that might help robots understand human mental states. Progress in emotion AI (such as through MIT spinout Affectiva) might enable robots to understand social cues like facial expression or intonation and build relationships with us in some meaningful way. Robot ethics is another area where we had better make some progress in the decade ahead, given that it might come in useful (Gent, 2020).

Energy storage is a major bottleneck for mobile robotics, because the battery packs are too heavy. If we found ways for robots to harness energy from their environment, or to provide massive, distributed energy sources upon demand, that would be one solution.

The future of synthetic biology

Synthetic biology is a new interdisciplinary area that involves the application of engineering principles to biology toward the design and construction of new biological entities such as enzymes, genetic circuits, and cells, or the redesign of existing biological systems. The synthetic biology industry has seen investment in excess of $12 billion, both from established biotech firms and venture capital, a third of which happened in the breakout year of 2018 (Cumbers, 2019).

The history of synthetic biology begins with the genomics revolution (with the dramatic increase in genomic sequence data and the exponentially lowered

DNA synthesis costs) and subsequent rise of systems biology in the 1990s, although the first international conference in the field was held at MIT in 2004. What made the difference was being able to apply engineering discipline to biological processes, ultimately starting to experiment with how to create, control, and program cellular behavior (Cameron et al., 2014). The metaphor in these formative years was the electric circuit, hence the notion of genetic circuitry as well as switching on and off gene expressions.

The impact of science and technology on synthetic biology

More than any other technology, synthetic biology challenges our assumptions about what it means to be human. Gene editing has the potential to change the stem line, which would alter our DNA, or nature's DNA, forever.

Some forms of synthetic biology, such as supplements, might be relatively harmless (and beneficial), while other forms (such as cloning humans or tinkering with viruses) might lead to humanity's extinction. The project on species revival, where Harvard's George Church wants to revive the mammoth and put it back in Siberia, would have to be one of them, given the debatable value of such a move. It could be seen as a type of origin fundamentalism, where he wishes us back to some origin point in the state of nature where the balance of power between humans, nature, and animals was more equally distributed (and humans struggled more).

The impact of regulation on synthetic biology

Regulation in the domain is in its infancy. The primary international forum deliberating the regulation of "synthetic biology" is the Convention on Biological Diversity (CBD) (https://www.cbd.int/), a multilateral treaty drafted in 1992, ratified by only 30 states, and a result of the Earth Summit in Rio, currently under oversight of the UN Environment Programme (UNEP). As with all such treaties, they are only as strong as the national implementation, which has happened with some force only in countries such as Singapore, the United Kingdom, New Zealand, and Tanzania, while the United States has not ratified the treaty.

For sure, the world, and particularly countries with strong scientific teams innovating at the edge of synthetic biology both in academia and industry, will need regulatory approaches that can meet the demand. However, the challenges are many. Not only do regulators have to keep up with the science in some way, but scientists need to learn more in order to even explain themselves, so to speak, especially when the prospect becomes creating entire novel synthetic organisms (Cameron et al., 2014). The real challenge starts when the newly created genetic circuits start becoming so complex and useful that other labs and application fields begin to build on the work of others. Having a framework for what is appropriate to share, and having some way to measure the risk of such sharing, become paramount in order to prevent runaway circuitry that escapes the lab or perhaps destroys the work in the application lab because its function is not entirely understood.

The European GMO Directive (2001/18/EC) put in place a moratorium on genetically modified organism (GMO) approvals, yet allowed for GMOs and food or feed made from GMOs to be marketed in or imported into the EU, provided that they are authorized after passing strict evaluation and safety assessment requirements (notably risk assessment, monitoring, labeling, and traceability) that are imposed on a case-by-case basis. The cultivation of GMOs, however, was left to individual member states (Law, 2020). Given the prohibitive nature and unclear adoption, the current regulatory framework has been criticized for not "proportionately, predictably, and enforceably" regulating organisms that have been bred by genome editing (Wasmer, 2019).

Companies and research institutions that employ new breeding techniques (notably deploying CRISPR/Cas9) face considerable legal uncertainty. In fact, the strict EU approach, while laudably cautionary, arguably restricts agricultural innovation and will likely have to be amended in some way in coming years.

In 2019, Chinese and Russian scientists conducted several human genome editing experiments despite an international call for a moratorium on human genome editing. China has arguably also deployed genetic targeting through face recognition surveillance systems ("learning race from a face") against Muslim minority populations, notably the Uighurs (Wee and Mozur, 2019). In the US, the Engineering Biology Research and Development Act of 2019 passed in the House and moved into the Senate for consideration where it has stalled and likely will past the 2020 presidential election and until well into 2021.

The impact of business models on synthetic biology

Emerging applications of synthetic biology include biosensing (sensors capable of picking up information from the biological environment), therapeutics, biofuels, pharmaceuticals, and novel biomaterials. Even though medical applications are likely to come first, synthetic biology will reshape the business models across industry sectors that can use synthetic materials in production (see Figure 6.4). Because of its potential to combine characteristics from nature (growth) with engineering capabilities (exactness, predictability), biodesign has the potential to reshape many of the valuable materials we know today. For example, synthetic biology is now being tested in the textile industry.

Currently, the global market is dominated by the US company Thermo Fisher Scientific, Inc., which offers a broad portfolio of synthetic biology products such as DNA fragments, software, engineering kits, DNA oligos, and clones, and has a vast distribution channel.

The growth of this market will depend on the demand for synthetic genes and synthetic cells, the growth in the range of applications of synthetic biology that get beyond the R&D stage, and the speed with which the cost of DNA sequencing and synthesizing continues to decline, the degree to which R&D funding continues to rise (government funding being a big question mark), and the success of industry-wide coordination initiatives in synthetic biology, and increasing investments in the startup space for synthetic biology (MarketsandMarkets, 2020).

Figure 6.4 Synthetic biology as a force of disruption

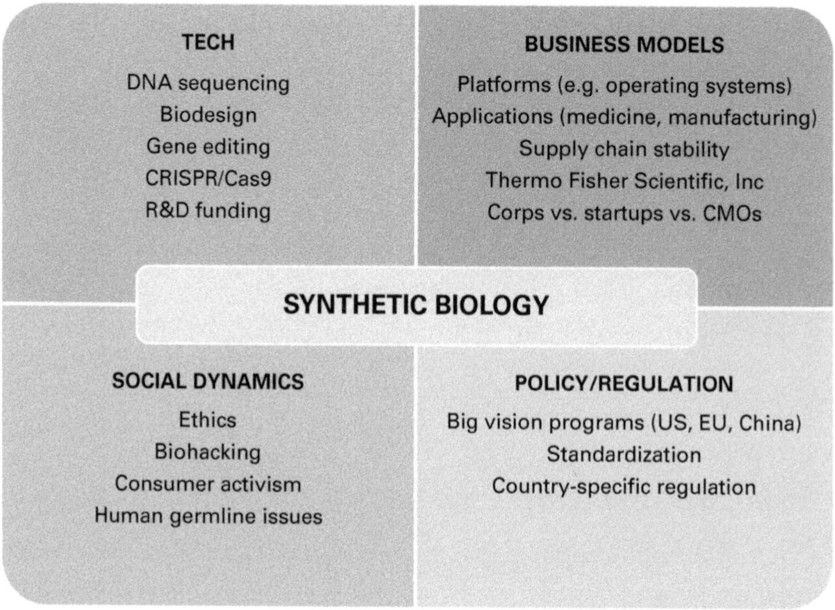

TECH	BUSINESS MODELS
DNA sequencing	Platforms (e.g. operating systems)
Biodesign	Applications (medicine, manufacturing)
Gene editing	Supply chain stability
CRISPR/Cas9	Thermo Fisher Scientific, Inc
R&D funding	Corps vs. startups vs. CMOs

SYNTHETIC BIOLOGY

SOCIAL DYNAMICS	POLICY/REGULATION
Ethics	Big vision programs (US, EU, China)
Biohacking	Standardization
Consumer activism	Country-specific regulation
Human germline issues	

Some of the key challenges to overcome before biodesign can enter the regular supply chain include standardization (including building an "operating system" akin to what we have in the software industry), scalability, and specialization in key industrial application areas (Cambridge Consultants, 2018). A big chunk of these challenges will likely fall on the next generation of contract management organizations (CMOs), given that manufacturers are more likely to trust them than smaller startups or bigger conglomerates.

Whether the new business models required to scale biodesign will appear within this decade or at the beginning of the next also depends on whether entrepreneurs come up with single profitable applications and are able to execute to market to sell that product. Scale will also depend on platform innovations, such as the aforementioned need for an "operating system" of sorts.

The impact of social dynamics on synthetic biology

The major influences on social acceptance seem to be "the state of knowledge and awareness of the benefits of the biotechnology; confidence and trust in the producers of the biotechnology; and the notions of risk, uncertainty, and complexity" (Pauwels, 2013).

There will be a heroic attempt by the industry to describe the synbio industry in the familiar terms of the computing industry, as a form of biocomputing. If that succeeds, adoption can likely happen more rapidly than if it fails, and the narrative becomes that this industry must be scrutinized as an entirely new thing that we need to be wary of and we must apply a cautionary approach not to destroy the "natural" ecosystem. The debate of what constitutes "natural" is highly dependent on cultural and political context and is more sensitive in Europe than in Asia and America, which operate largely in a "post-nature" paradigm.

Important debates about the increasing adoption of biodesign should be happening across all domains. While its use in industrial materials isn't of great concern, its use in areas such as genome editing or ingestion, directly or indirectly, would require more careful scrutiny. Developing a deep perspective on the ethics and acceptable practices within this industry is a worthwhile investment. Biodesign will, for better or for worse, start shaping our decade, and in some ways, it already is.

What synthetic biology is capable of today

Today, synthetic biology is capable of using the basic building blocks of nature to craft products or improve basic products. It started with yeasts and is now moving up the organism food chain, albeit slowly, and a good bit of time has been spent on the genetic circuitry of the *E. Coli* bacteria, which almost has the same status in molecular biology as the mouse has in experimental biotech and psychology.

One of the best ways to track current use of synthetic biology is through BIO, the world's largest trade association representing biotechnology companies, academic institutions, state biotechnology centers, and related organizations across the United States and in some 30 other nations (https://www.bio.org/). Current use cases cited by BIO (2020) include forms of naturally replicating synthetic rubber (as opposed to degrading a finite resource such as the rubber tree), renewable bio-acrylic (to reduce dependency on oil), green chemicals (reducing agricultural waste), custom genes for vaccine development (speeding it up drastically), synthetic antibiotics production (speeding up the production process), biofuels using low-cost sugars (replacing petroleum), synthetic fermentation of adipic acid, a commonly used chemical in the production of nylon (reducing environmental footprint), and bioplastics (through naturally fermented polymers).

What synthetic biology might achieve in 10 years

Over the next decade, synthetic biology will make considerable advances, and depending on regulatory oversight, might develop approaches that have the potential to change the direction of evolution in fairly drastic ways, hopefully positive, although there are no blanket guarantees. The field is certainly well positioned to advance medicine over the next decade through next-generation diagnostics and gene and cell therapies. While cloning certainly could become a key issue this decade, more likely the narrower therapies that target truly devastating human diseases like Alzheimer's are more likely to dominate the focus.

Beyond transforming biotechnology and medicine, it is quite clear that synthetic biology will have significant impact on manufacturing across the board, particularly upon consumer goods (particularly the beauty market) in the first instance.

Across the world, GMOs will rescue plants and wildlife that are struggling under conditions of rapid climate change. This will become much-needed help for farmers and fishermen and consumers alike. IVF clinics will tweak their approaches, achieving healthier offspring. Gene editing will be attempted in embryos, fetuses, children, and sick adults as a therapeutic modality of choice for the elite or those living in single-payer, government-sponsored health systems. The field will likely revolutionize vaccine development, especially if the current mRNA vaccines for COVID-19 do get approved and don't engender any observable cataclysms. Artificial blood might be on the horizon. New life forms will be "created" or created out of hybrids. Telemedicine might become a true alternative for certain procedures (Caplan, 2020).

One might hope to reduce the incidence of rare genetic disease. There should be some progress in virus-resistant plants and algae and temperature-resistant plant and wildlife capable of better dealing with manmade climate change. The advances in research on aging should also be impacted. The intersections with cognitive science will become more apparent and there might be a true fusion between the field and strands of computer science, given the experiments in brain–machine systems. Mitigating all forms of environmental degradation will definitely still be a worthy challenge. The ability to respond quicker to epidemics and devise better therapies will become both necessary and immensely useful. There might also be a role to play in enabling further space exploration.

Consider precision medicine, one of the heralded super-therapy strategies of the decade we are entering. The notion that we could be so precise that we could develop medicines for very specific groups of individuals, even for one individual, and highly specific conditions and circumstances (for instance, five years old, African, female, with preexisting conditions and mutations in genes TP53, APOE, and TGFBI) has been a pipe dream in medicine in some time. Most therapies are at the research stage at present, but some experimental therapies are entering the larger research hospitals.

Precision medicine use cases include cancer treatment, preventative healthcare, drug efficiency, reducing drug side effects, improving clinical trials, and eliminating trial and error.

Rehabilitative medicine might be able to repair rather than compensate through neurostimulation, wearable robotics, and gene repair, perhaps with the use of universal donor cells that are reengineered. Artificial blood based on human–pig chimeras enabling efficient transfusions and transplantations even for rare blood types is another barrier that might be broken. Finally, tying into the discussion on 3D printing, the ability to print living cells and perhaps organs for clinical use is also on the horizon (Brownell, 2020).

Whatever happens, one thing is certain: the benefits (and drawbacks) of synthetic biology will extend far beyond healthcare and agriculture, which means it cannot be viewed as a specialty field. Rather, it is more akin to digitalization—it affects all of us in society-altering ways.

Synthetic biology is moving so fast right now that rather than spend too much time rehashing what has been done in the past (meaning, in the past decade), you need to look as deeply as you can to the ongoing experiments charting the near future (meaning, the next three years). That is why, more than in any other field, tracking the innovators is the only approach that will yield any kind of understanding of the evolving dynamics.

The future of 3D printing

3D printing is a new approach to manufacturing products where instead of making a mold, as would traditionally be needed for plastic or metal parts, you create separate layers of a product and mold them together layer by layer. 3D printing is also referred to as "direct manufacturing" since you manufacture without going via molds, or even "additive manufacturing" since you are adding ingredients until you get the desired product as opposed to cut-off parts which leads to waste.

The process of additive manufacturing grew out of commercializing the stereolithograph approach (SLA) to layer-by-layer printing, whereby liquid plastic is converted into solid objects, back in 1987 and it has blossomed into a thriving, wide-ranging 3D printing industry and a bunch of suppliers. An extension, digital light processing (DLP), deploys light to quickly harden the end product, which makes it a bit more practical. That development followed decades of experiments in additive manufacturing, which is perhaps a better name for the overall field.

Japanese manufacturers (Sony, Mitsui, etc.) concentrated on SLA but in the US and Israel other approaches have flourished, notably FDM (fused deposition modeling) printing, which is by far the most common inexpensive method of 3D printing, and one taken by MIT spinout Formlabs. With that approach you can use the scaffolding of a support material for the object being formed during the printing process which can later be removed or dissolved.

One challenge with SLA is that the main ingredient, resin, is available in a only very limited number of colors, generally a clear or translucent material and white. In contrast, there is a rainbow of colors available for FDM printing. A third method, laminated object manufacturing (LOM), takes thin materials like paper or plastic sheets, cuts them to a specific shape, and then uses adhesive to glue one layer to the next, which is perfect for rapid prototyping of large objects in architecture or construction where a chemical process might be overkill in terms of cost, sustainability, waste, or complexity. On the high end, you find selective laser sintering (SLS), which can make use of any material that is capable of being powdered and fused with heat, including thermoplastics, steel, aluminum, titanium, and other metals and alloys. A variant, selective laser melting (SLM) involves fully melting the powdered material.

Throughout the last decade, the material inputs have extended from plastics to liquid form metals, which has spawned a burgeoning metal 3D printing process that promises radical breakthroughs in aerospace engineering products. The first parts have been produced in space.

The impact of regulation on 3D printing

In the US, the first guideline specifically governing the additive manufacturing of medical devices was published in December 2017 by the FDA. Even though, according to the European Commission, by 2021 the 3D printing market could be worth €9.6 billion, in the European Union there are still no specific guidelines for the additive manufacturing of medical devices or pharmaceutical products (Miglierini, 2018), which creates regulatory uncertainty. There is, however, strong R&D support for innovation in the domain.

Could invasive, patient-specific devices such as implants and prosthetics be customized to suit the recipient's anatomy? Could surgical instruments and other pieces of medical equipment be altered—and perfected—based on feedback from healthcare professionals? It's possible, but it would require careful regulatory scrutiny.

The impact of business models on 3D printing

3D printing eliminates the long lead times associated with conventional mass production. In 3D printing, spare parts could be printed rapidly on location or at home, and with much quicker turnaround times by the manufacturer.

The on-demand production made possible through 3D printing allows companies to satisfy the demand without the high costs of custom or mass production in their factories. Instead, you download their specifications (the "electronic blueprints") and print it yourself on site.

At best, 3D printing enables a co-creating process between customers and companies, resulting in a made-to-order, customized product. Both parties can exchange prototypes and do test runs, which enables rapid prototyping. Because of this new distributed nature of product development there is also the possibility of crowdsourcing both the design ideas and other parts of the production process.

In the consumer segment, 3D printing allows digital manufacturing, home fabrication, and mass customization without the traditional overhead. You can change colors or even base materials without much pain. Three business models are emerging in parallel: product-based, platform-based, and combined models (Rong et al., 2018).

While the allure of 3D printing is its convenience, capturing value can become challenging because so much of the production process is outsourced to the customer. If open sourcing specifications become the norm, there is the possibility that most of the value (in some products) will reside with the material as well as 3D printer hardware manufacturers, as opposed to with those creating the electronic blueprints. On the other hand, complex blueprints should stand a good chance of being proprietary and would be covered by patents, so they would require licensing to deploy at scale.

The question will be who will bear the costs throughout the value chain, from the product development through to the material cost, the supply chain costs (materials will have to be shipped to the manufacturer, to the end client, or some combination of the two), the hardware cost (complex 3D printers are costly), the maintenance cost (such machinery may break down and require maintenance), as well as the upgrade cost (new models appear all the time). Such issues are shared with other products such as computer equipment. Rent, lease, or buy decisions and models will likely co-exist for some time. The question is also one of convenience. Just because you can assemble a computer on your own doesn't mean the average consumer is interested in learning how to do it and actually doing it. For example, my son and I recently opted to buy a gaming PC from a manufacturer even though we had spent three months researching build-it-yourself videos. Assuming that affordable, high-definition, multimaterial personal 3D printers become available, what percentage of consumers will make use of that option instead of ordering from a store, and for which products (Rayna and Striukova, 2016)?

The impact of social dynamics on 3D printing

Many of the social dynamics of 3D printing are still largely unknown and will depend on which business models pan out. However, it seems clear that the promise of community-based innovation is one very real possibility. Crowdsourcing innovation virtually at a grand scale would also vastly change the way individuals interact with industry, with product innovation, and might alter the way products become embedded in society more generally. If that happens, we should see more products that work better for their users, fewer product failures, and products better tailored to specific social groups and use cases.

The fact that the production process is radically different with 3D printing has numerous implications for the social interaction of workers on the shop floor. It offers the potential that interaction between workers can be better understood because it is captured digitally, which could improve both the production process and the social relations between workers and those between workers and end users.

We are just beginning to see the adoption of 3D printing in local communities, cultures, and social movements (see Figure 6.5). The most media-hyped scary scenario might be that of 3D printed guns. However, consider the positive potential of 3D-printed medical gear, prosthetics, vaccines, low-cost housing, and anything needed for disaster relief, and you can start to imagine new social dynamics fostered by this technology. Ultimately, with less transport of goods, that would mean less pollution, although the materials going into 3D printers still have to be produced somewhere and brought to the printing site. Also, the printing process itself is highly energy intensive.

What 3D printing is capable of today

Today, most 3D printers are capable of printing plastics (ABS, PLA, nylon), but metal alloys, ceramics, wood particles, salt, and even sugar and chocolate can be used to create 3D-printed products (Rayna and Striukova, 2016). Up until a few years ago, most printers could print with only one material at a time, but that is changing rapidly for high-end models and will also change for smaller desktop models, making them readily available to prosumers.

The range of products that can be produced is wide and rapidly expanding, too, including "prototypes, parts, molds, tools, body parts (organs), prosthetics, toys, art, food items, musical instruments, furniture, and clothes" (Rayna and Striukova, 2016).

The entry point pricing for 3D printers is down to a few hundred dollars, making them attainable for most consumer segments. They are increasingly available at regular department stores. Obviously, high-end models that can print bigger, multimaterial products can become quite expensive. For example, in 2017, the Desktop Metal Studio system cost $120,000, but offices or

Figure 6.5 3D printing as a force of disruption

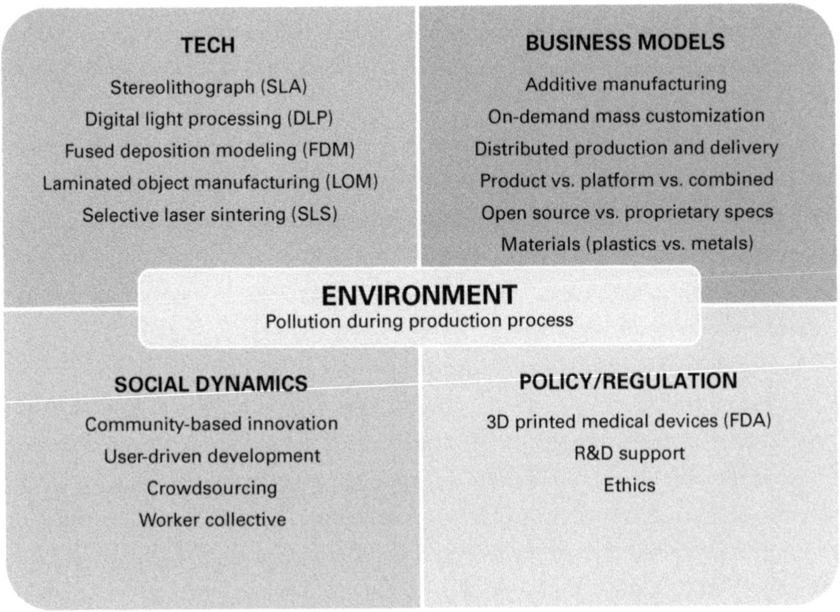

TECH

Stereolithograph (SLA)
Digital light processing (DLP)
Fused deposition modeling (FDM)
Laminated object manufacturing (LOM)
Selective laser sintering (SLS)

BUSINESS MODELS

Additive manufacturing
On-demand mass customization
Distributed production and delivery
Product vs. platform vs. combined
Open source vs. proprietary specs
Materials (plastics vs. metals)

ENVIRONMENT
Pollution during production process

SOCIAL DYNAMICS

Community-based innovation
User-driven development
Crowdsourcing
Worker collective

POLICY/REGULATION

3D printed medical devices (FDA)
R&D support
Ethics

labs could also rent the system for $3,250 a month and a production system would set you back $420,000 (Kolodny, 2017). Arguably, prices have slowly and slightly been decreasing, although Desktop Metal's system seems to cost exactly the same in 2020 (Cherdo, 2020). Metal printing systems today generally range from $50,000 and $1 million depending on their complexity. The machine itself often represents only a fraction of the overall cost, considering that metal powder can cost between $300 and $600 per kilogram (Gregurić, 2020). Most metal printers today are not yet plug-and-play machines and must either be operated by or supervised by trained workers.

Wohlers Report (2020), the most read industry guide, documents more than 250 applications of additive manufacturing today. One company to watch is the American-Israeli Stratasys (founded in 1989), the world's largest 3D printer manufacturer, selling to the aerospace, automotive, and medical industries. Judging from Stratasys' clients, 3D printing is most widely diffused in the US, which tops the rankings, followed by Japan, Germany, China, and the United Kingdom. Other key players in the industry are 3D Systems, the runner-up, as well as GE, Intel, Home Depot, Autodesk, and Amazon, as well as a growing body of startups, notably Boston-based Desktop Metal, Formlabs, and Markforged.

Metal 3D printing has been used to create objects ranging from airplane fuel nozzles to hip implants. However, items made using metal 3D printing are susceptible to cracks and flaws due to the large thermal gradients inherent in the process. For that reason, so far, 3D printers have mainly been used to make prototypes in the manufacturing industry. Nevertheless, with recent advances and wide interest, they have started to be used to create jigs and parts for product manufacturing or even for end products, paving the way to deploy 3D printers for mass production. Incidentally, the solution to these quality issues is another technology, AI. Experimental methods now are trying to deploy advanced simulation models to detect and correct such flaws as they happen.

Prosthetics is one area where 3D printing is already implemented. During COVID-19 it has also been used to print personal protective equipment to complement a failing supply chain (Bell, 2020). Prototyping new treatments and devices is another use case.

What 3D printing might achieve in 10 years

The next decade promises 3D printing with new materials, creating new hybrid materials, perhaps, including cellulose, the world's most abundant polymer, as well as carbon nanotubes. Such nanomaterials could be used in flexible electronics and low-cost radio frequency tags. Medicine is next. Going into biological material might also entail growing human tissue, both as replacement organs and as a testing ground for medical experimentation relevant for drug discovery and related effects, potentially even growing brain organoids that could unlock the secrets of Alzheimer's disease (O'Leary, 2019). Medical devices is another promising application. Having said that, one of the broader benefits of 3D printing is that it is an enabler for a more digitalized supply chain, which could transform fabrication and production.

Astronauts in space have had a basic 3D printer since 2014, but in 2019 a new version appeared, The Refabricator, built specifically to enable the use of recycled parts, something traditional 3D printers cannot typically do (Goldsberry, 2019). 3D printing lets astronauts create parts quickly, currently in a matter of a few days, which is not helpful for immediate problems but certainly better than waiting until you can ship a part into space given that the time between cargo flights to the International Space Station (ISS) is 3–4 months at least. One day the technology might be used to print habitats or landing pads on other planets.

Dutch startup MX3D has developed a metal 3D printing system that can print metal parts without supports, by combining a multi-axis robotic arm with a welding machine. With this method, metal wire is pushed through a feed nozzle where it is melted by an electric arc and successively added onto the build platform. The company attracted attention by building a metal bridge that way, in a process that took a year, as well as printing a futuristic bike, which took only a day.

Taking a different approach is MIT spinoff Desktop Metal, which designed the first office-friendly metal 3D printing technology.

Conclusion

Throughout this chapter I have argued that the five technologies that matter the most at this moment in time are artificial intelligence, blockchain, robotics, synthetic biology, and 3D printing. While I could think of at least ten other candidates to be included in the list, I selected these five technologies because they interact with each other, creating previously unthinkable conditions of technological, biological, material, social, and psychological change.

Consider how 3D printing makes use of AI-enabled machine learning, which enables further digitalized supply chains, which together with the increased use of robots throughout that supply chain, contribute to automating the manufacturing industry, which has ramifications for every other industry. When blockchain is used in manufacturing, it makes the supply chain more secure so you can verify product origin, integrity, and whereabouts, and enable smart contracts that change the business dynamics, too. Synthetic biology also reshapes manufacturing because the products we know might now take on biological properties we had never conceived of, literally growing products instead of, or in addition to, making or printing them, bringing a plethora of new products to market.

These technologies together foster entirely new ways of being ourselves, of being designers (in a design-thinking sense that we are all aspiring designers of worlds), citizens, consumers, financial actors, parents, lovers, in fact across all or most of our multifaceted roles and identities.

However, it is important to remember that the successful adoption of technologies depends on regulation, business models, and social dynamics. Regulation cannot happen in a vacuum. Business models can literally sweep away technologies and foster others. Social dynamics can make one technology take off and may doom another to failure, they can foster the need for regulation or the impetus for innovation. As we have seen with COVID-19, the contextual factor of the environment can suddenly, and without notice, move to the foreground. This could happen again with another virus, or perhaps with a natural catastrophe, or perhaps with ecological degradation caused by climate change or interrelated phenomena. Foresight does not mean seeing the future but acting on possible futures, much like being studying to become a martial art expert does not mean you necessarily seek out street fights (it is quite the opposite, in fact).

No matter what we might think of the mixed blessings of AI, or its current limitations, it will fuel innovation across industries, in government, and in civil society. Blockchain is not only slowly emerging as the next financial platform to follow the invention of money. It is also a possible answer to the trust issues plaguing most sectors of our society: verifying the provenance, identity, supply chain, and trajectory of people, products, and content is among the main pathways I currently see to recreate that trust. With robotics going mainstream, humanity can finally get a respite from having to be exposed in dangerous, repetitive, or contagious scenarios, and we can achieve most-anticipated efficiencies in our work and, perhaps, even in leisure. Synthetic biology will undoubtedly become a viable path to greater control with nature and more efficient engineering of our surroundings by switching on desirable biological characteristics in order to enable new materials, yet with the efficiency far surpassing current technology tools. And 3D printing, with its promise of decentralized production, near instant supply, and remote consumption of advanced products assembled through additive manufacturing processes that will become cheaper and simpler and ever more capable, will ensure that the last mile problem is less of an issue toward the next decade. This, in turn, if we set it up well, will potentially enable us to correct for inequalities and drawbacks stemming from location and switching on a proximity-enhancing layer that may come in quite handy. If there is one more technology I wanted to mention, it would be augmented reality.

To be future-proof we cannot limit ourselves to only picking a few technologies to watch. Rather, it is essential to become a deep expert in one domain and to be well versed in dozens of other domains so that you can communicate across and have the absorption capacity for all kinds of novelty. This goes far beyond the currently recommended approach of becoming a T-shaped expert (expert in one and shallow awareness in others). It also has to do with developing the teeth to be curious but with a bite—so you can still be critical to new perspectives coming from fields where you have not spent a decade preparing the ground.

Key takeaways and reflections

1   Pick one of the five technologies and map your knowledge of them in the following way (ideally you do so before you read the chapter, after you read it, and then after two weeks once you have brushed up on the topic).

2   Map the relationships between the five technologies using a blank piece of paper and some arrows. You can deploy the five forces of disruption, the subforces discussed in this chapter as well as any subforces you find relevant.

3   Imagine what the next five technologies that matter might be. Jot down what you know and believe about each of them. Compare your notes with somebody you trust and discuss commonalities and differences in your assessment.

How to become 07
a postmodern
polymath

In this chapter, I outline the growth opportunities for individuals who embrace the landscape created by new technologies interacting with other disruptive forces—specifically, how to become a "postmodern polymath," meaning an independent thinker capable of catalyzing change and innovation through combining technology insight with other domains. The implications from looking at five forces of disruption (tech, policy, business, social dynamics, environment) would be that some combo of each needs to be deeply understood. Beyond that, extending the typical notion of T-shaped or Pi-shaped expertise away from the notion of expertise to capture systemic understanding, ambitiously aiming for wisdom, is what's truly needed. I will look at practical approaches to aim for this type of polymath-empowered systemic wisdom at individual, family, and societal levels.

T-shaped and beyond

In a now famous 2010 interview with *Chief Executive* magazine, design firm IDEO's CEO Tim Brown popularized the term "a T-shaped person" as somebody with a depth of skill that allows them to contribute to the creative process (the vertical stroke) which in the IDEO context may stem from knowledge of any number of different fields: an industrial designer, an architect, a social scientist, a business specialist, or a mechanical engineer. In addition, such a person would have the disposition for collaboration across disciplines (the horizontal stroke). Arguably, the combination of those two aspects produces the empathy and enthusiasm needed for the collaborative work that characterizes the contemporary workplace. The argument is straightforward and is both better branding and a more accurate description of best efforts to combine knowledge than decades of talking about interdisciplinary or cross-disciplinary efforts in academia. However, the notion also comes with its own challenges. How do you prove that this is better? How much breadth is enough? How much depth? How much better is it than the alternative? How do you hire for it? How do you become a T-shaped expert? Is T-shape enough?

For sure, the traditional educational system is built to produce I-shaped people at best, meaning people skilled in one domain (a profession, an applied vocational skill, a language, or one of the arts). Taking T-shaped as a slightly more ambitious ideal, the credo is to become an expert in two or more domains and with broad expertise so you can communicate with a wide variety of people. In strategic HR they have started to take the metaphor of the "T-shaped expert" as something of an ideal, perhaps because it focuses selection efforts away from simply testable domain skills and into combinatory skills, which are vastly more useful but harder to test for.

In reality, becoming a T-shaped expert is perhaps possible to learn, and recruit for, but is clearly not enough to become a shaper of future technology. In this chapter, I will go through how you should aim to become an expert in two or more domains (effectively a Pi-shape) and should simultaneously become deeply immersed in a dozen. However, in the process you need to move out of the "expert" attitude and start embracing commonalities, discovering forks in the road that might hamper progress, look for communication points, and start to truly combine these areas in practical (and theoretical) applications.

Beginning a journey of combining insights from several domains

Having a T-shaped profile is not enough. Even if you just look at today's innovators, they don't limit themselves to being experts in one domain. Rather, as we will see, most of today's innovations, as well as those that stand the test of time from the history of science, technology, and the arts, are *combinations* of expertise with a novel application either into a domain or creating a new domain altogether through an invention. Getting into

T-shape, much like managing to run three miles without stopping, is the preamble.

Transdisciplinary collaboration ability

Tim Brown was a precursor in the domain of business, but he did not invent the term or the practice of cross-disciplinary collaboration. There have been interdisciplinary departments at European universities for decades. In the US, bachelor's degrees in interdisciplinary studies programs have also existed for decades, although their popularity has gone in waves and some have shut down. One of the challenges in an educational context is synthesis, how to truly reconcile the insight from different domains into a holistic understanding. This is not a simple problem, and the source of the problem resides deep in epistemology, the theory of knowledge.

Each domain is often incommensurable with others, and as we know from the history of science, even paradigms within the same science may have this characteristic. Science theorist Thomas Kuhn (1962) famously claimed paradigms only died with their founders and was pessimistic on how to solve the issue.

Here are some distinctions to keep in mind that show how polymaths become gradually more complex. First, *interdisciplinary* implies interaction with at least another domain, which is the first step. You need to find learning modes and material that are open to this possibility, or you need to think carefully about these interactions yourselves. Second, *pluri-disciplinary* is often used about applied topics of policy, such as health, and implies domains are not just interacting but also related in some important fashion. Pluri-disciplinary means that it is universally recognized that input from a few specific domains are needed to fully understand the issue, but it does not mean you necessarily need to have those domains understood by the same person. Third, *cross-disciplinary* is the juxtaposition of two domains but where boundaries are occasionally crossed and there is specific outreach between experts but typically no integration of the domains. One example might be the early efforts in robotics, where engineering was combined with computer systems. Fourth, *multidisciplinary* means juxtaposing more than two domains yet without attempting to integrate the domains as such. The best example here would be today's robotics efforts, which build on insight from experts in engineering, mathematics, informatics and computer science, neuroscience, or psychology, working together in some fashion. Another example might be bioengineering, where the best elements of biology and engineering are combined.

The next ambition level would be *transdisciplinary*, which involves common axioms, that is finding the core ideas in common between domains. Early examples of transdisciplinary approaches include Marxism, structuralism, and feminist theory, but the best contemporary example might be systems theory, which builds on a combination of mathematics, psychology, biology, game theory, and social network analysis (and sometimes more). Systems theory is used to understand organizational change and complex phenomena where many domains, people, and systems interact, such as climate change.

With transdisciplinary approaches, the imperative is to develop a unity of knowledge, which is at times insurmountable, but which would yield vast improvements in current understanding if/when it succeeds. The best current examples of such efforts are synthetic biology, which specifically combines biology and technology's best efforts (see Chapter 5), and artificial intelligence, as it attempts to unify the knowledge gained in psychology's study of cognition with applied neuroscience, with computer science and algorithmic thinking derived from statistics and mathematics more broadly.

As you look more closely at any of the emerging technologies mentioned in Chapter 2, you will find that at the highest level of theoretical or practical application, and even when crafting products based on them, although partial success might be had with far less ambitious integration of knowledge, a transdisciplinary mindset is often needed to make true progress.

If you want to learn about systems theory, there are many blogs, books, podcasts, and online courses dedicated to the topic (see appendix). There was an early attempt by the OECD (1972) to clarify these concepts, but the report remained in circulation only in specialist policy circles instead of being widely understood in the population, which would have made a great difference in our educational system over the past 50 years. Changing such perspectives historically takes a long time, although change has now arguably accelerated to a point where if the above is not fully clear to you, you are falling behind already.

When you think about *disciplinary boundaries* it is important not just to have academia in mind. The tensions (a) between theory and practice, (b) between research and teaching, (c) within and between academic disciplines (qualitative vs. qualitative methods), (d) between scientific findings and commercial application, as well as (e) in practical discussions about a topic you are trying to solve during product development or in the workplace more generally are all important.

The transdisciplinary individual

In a 2014 article, Augsburg outlines four dimensions that transdisciplinary individuals have: (a) an appreciation of an array of skills, characteristics, and personality traits aligned with a transdisciplinary attitude; (b) acceptance of the idea that transdisciplinary individuals are intellectual risk takers and institutional transgressors; (c) insights into the nuances of transdisciplinary practice and attendant virtues; (d) a respect for the role of creative inquiry, cultural diversity, and cultural relativism.

One of the challenges with transdisciplinarity is to get beyond disciplinary science's insistence that while the methods of inquiry could be (somewhat) shared, the axioms are not.

The greater aim—adding value to society

The best way to add value to society is through attempting to bridge disciplines and build a transdisciplinary perspective where you deeply integrate the concerns and axioms of the domains you consider and apply a unified lens.

Whether you work in a startup, are involved in innovation at a large corporation, work in government trying to create good policy initiatives, or are trying to be a good parent, a transdisciplinary mindset is what's required to succeed in a way that adds value to others (and beyond your own immediate benefit).

After all, this is likely why evidence-based policy making is so difficult, for instance, because scientific knowledge must be combined with local knowledge and concerns, leading to a more effective interface between science, policy, and society. Similarly, successful product design involves merging the concerns of various domains needed to build the product with the constraints (and opportunities) of the use context and target market. Imagine trying to create a product while claiming there is no way to merge the designer's visions with the technologist's prototype. It cannot work. Similarly, in teaching and learning, both the teacher and the learner must "merge" with their media (the screen, the software, the books, the voices, the emails, the chats), forging intense energy through fusing the curriculum, the medium, the mentor, and the mentee in a "dance" (Subrahmanian and Reich, 2020).

Becoming a polymath

The way to start training for a transdisciplinary mindset is to pursue poly-maths. You should not pursue knowledge in one domain for its own sake. What we are looking for lies beyond the disciplines, since domains of knowledge are merely the tools (or the *techne*, if you will) for our decision making, product development, and attempt to induce progress.

The challenge is that the notion of transdisciplinarity introduces new concerns that are born from the attempt to merge traditions, as well as from concerns in society (such as sustainability, ethics, effects on the community) and other global issues. However, the results are often transgressive, as in they might violate the boundaries, challenge social norms, established truths, elites, and even ethical constraints. Just as science is best conceived as an activity that constantly surpasses, erases, and rebuilds all kinds of boundaries, either disciplinary, socio-ethical or ecological (Krings et al., 2016), and progresses mostly through failure and through analyzing failure (Firestein, 2016), such is also the quest of the entire practice of technological innovation. You should aim to design clever ways to experiment with things that seem related to what you are working on and should be *excited* when you fail because it now opens up a learning moment. Sharing your observations with others, reflecting on what happened, and the psychological adaption that happens as you attempt to recover, perhaps from a costly failure (in time, money, or reputation), is what, ultimately, will put you in a position to make a pathbreaking discovery one of the next times you try.

Success is very often nothing more than a rescue of an impending failure at the exact moment before it would have failed. To achieve this, two things are required in addition to a polymath mindset: stamina and utter self-confidence. Those are much harder to learn, actually, and are either innate or learned slowly through the first few decades of your life as you "grow up" and socialize into your own specific sociopsychological identity as a mix of heredity, personality, circumstance, and daily effort, and some luck or lack thereof.

Designing your own polymathic learning path by seeking mentorship

In reality, achieving transdisciplinarity is even more complex, since different teaching philosophies also draw to a varying degree on insight from different disciplines. The *action learning* (also called experiential learning) movement

(Kolb, 1985), for example, vehemently argues you cannot learn without practicing a trade, or at least practicing while you learn, even if the domain is an academic subject with a lot of theory to learn as well. Imagine, then, that the "practice" domain is an additional disciplinary variable.

Apprenticeship learning is an even more traditional answer to this problem where the notion is that without an apprentice learning directly from a master or expert, learning by doing and getting critiqued, ideally with a small group of students around, true insight cannot transfer properly. Because it induces reflection, it has been proposed as a great way to learn a professional practice as well (Schön, 1983).

However, what that tradition has ignored is the highly personalized bias that a single "master" introduces into the learning process, which may complicate learning for individuals who may have different needs. You need to pick your mentors with care. If you discover they are not the right mentors for you, you must move 1,000 miles away from them because a bad mentor can set you back years.

Science does not always have too much to say about creativity since it is hard to describe. It is therefore not surprising that it is IDEO that takes the role as popularizer of the term *T-shaped*. After all, it is in the practice of design, especially the very broad definition used by IDEO, that the T-shape really gets the most mileage, since design as an engineering process is deeply steeped in cultural meaning (Subrahmanian and Reich, 2020). The only value of the T-concept that I can see is that it metaphorically can symbolize sharpness and edge, as well as stand for "technology," I guess. The T has remained with us because it has the power of a metaphor. T symbolizes tau, derived from the Semitic letters taw (ת, ࠈ, ت) via the Greek letter τ (tau), the 19th letter of the Greek alphabet. In Franciscan tradition it is a symbol of the redemption and of the cross. Given where society is going, you definitely will want to include technology within your own path to becoming a polymath. But not in the immediately obvious way that you may first think of, as in "learning how to code" or "learning about the periodic table." Here's why.

Technology is highly relevant to the T-shaped discussion because, arguably, technology is a method of accelerating the process of learning and application as it changes the environment around us more generally. In modern Greek, the word *techne* means to listen and the ancient term it derives from is often translated as craftsmanship. We need to "listen" to our technologies (as much as listening to our mentors) to use them well, and they each have a slightly different "sound" and will produce different sound based on who is activating them.

Having recently embraced a hobby as a podcaster, I find the metaphor of listening very useful. Audio is a direct channel into the soul, and by the way, a fairly direct way into our brain (through our ears, that is), which is relevant to the neuroscience of learning. In my busiest moments at university, I actively used autogenic training, a desensitization-relaxation technique developed by the German psychiatrist Johannes Heinrich Schultz by which a psycho-physiologically determined relaxation response is obtained. The method teaches your body to respond to your verbal commands. These commands "tell" your body to relax and help control breathing, blood pressure, heartbeat, and body temperature. However, I was also able to use the method to learn academic subjects at a deeper level. By inducing a self-hypnotic state, I could (and still can) meditate over complex topics whenever I was at a crucial point in my learning cycle (at the moment when a breakthrough is needed). I explore some further insights from current scientific progress in multisensory learning in Chapter 9 (merging with machines).

The approach is somewhat related to the relatively mysterious biodynamics for brain integration, which is a holistic therapy related to applied kinesiology I have also tried with some success. While these processes are still poorly understood by conventional science, I think it is well within reason to speculate that once we make progress in describing these processes and develop solid, evidence-based experimental approaches that are capable of intervening at the exact moment when our brain is actively "learning" or connecting separate domains (whether through invasive or non-invasive approaches), we will be able to make massive progress not only in the understanding of neuroplasticity but also in learning to control it better and achieve learning effects and, potentially, unlock human enhancement.

At this point, all I can say is that almost anything that you find has the potential to help you focus your energy, concentrate, or relax (or any of the above), whether it is exercise, meditation, autogenous training, specific activities (reading, sitting in a coffee shop, getting or giving a massage, conversing with others), and you should methodically explore how you could optimize and systematize such activity for greater results over your life span.

Picking the core set of technologies to specialize in

The question you must ask yourself at this point in your reflection is this: Which technologies are the most relevant for you? In other words, taking all into account—the relevance in society, your innate skillsets and previous experiences, your current interests, and commercial opportunities, all of it— what stands out?

The way to think of it is: you would want to be a unique contributor to *at least one specific tech-based advance* throughout the next decade. Even if we reduced it to the five technologies we analyzed in the previous chapter, if you had to pick one vertical dimension, which would it be? Would it be AI? Or would it rather be synthetic biology? Even more importantly, it is not enough to pick a tech domain. What would you want that contribution to be? Without being specific, you have nothing to aim for.

And if you had to pick one or two horizontal dimensions, going for the Pi-shaped profile, would it necessarily be empathy or teamwork, which seem to be the killer skillsets on the "soft side" of the equation? The answer must—controversially—be a resounding NO. The reason is that we should not be trying to optimize as individuals but trying to fit into productive teams, as individual insight (regardless of IQ or knowledge network connection) rarely can compete these days.

We also have to envision bringing in complementary skills to handle our own lives, much like the way we don't pretend to be HVAC, plumbing, electricity, gardening, or teaching experts. As much as we need empathy specialists, we also need leadership experts, problem-solving experts, communication experts, organization experts, marketing, project management, writing, creatives, extremely responsible folks, and you can go down the list of so-called soft skills, none of which is particularly soft at its peak levels of performance (who would claim that a jazz improviser on stage in a jazz club, or an author who just won a Pulitzer or a Nobel, or a four-language practitioner is practicing a soft skill?). The point is this, to make an impact, you need to stand out at *something*. It may matter less what that something is. What matters is your process of figuring out what your standout is going to be. Even if it takes you several years, you cannot excel without picking something. You just won't have the depth required.

Is the broad awareness only about soft skills like empathy and teamwork skills? Is the point of depth only to become an expert (or is there something generic about the mindset you learn when you dive deep)? I'd like to stop the discussion right there: The T-shape logic is not fruitful if we simplify that way.

The only way we can be T-shaped in the next decade is by having a broad understanding of most rapidly emerging fields (the horizontal axis) coupled with deep insight in at least two domains (so we can have grounds for recombining insight at some deeper level. One may refer to this as the TT-shape, although some refer to it as Pi-shaped skills or ∏).

This can get somewhat complicated but there's nothing magical about focusing on more than one discipline. The magic happens when you are highly attuned to which domains to specialize in and how to combine those domains in a unique way that yields value to your pursuits and to society. However, just like the mix of blue and green is the color cyan, we have to be prepared that the end result indeed is a mix of the concerns of each domain, and we cannot cry so much over which domain dominates in any individual product decision or societal design.

T-shaped problem-solving at McKinsey is, arguably, also about the way you approach a problem, ideally first by looking at the broader effects and then only when you know here to dive deep, do you do so (Working With McKinsey, 2013).

Reevaluating horizontal skills

In the real world, horizontal skills and vertical skills are not clearly delineated. For example, if we were to consider team skills, there is nothing that precludes that topic from being a vertical skill. It is not going to be enough to practice team skills. What is ultimately needed is deep, research-based reflection on what a good team is, how to foster its emergence, create conditions for it to flourish and thrive. Knowledge about these matters belongs to the domain of social psychology and is aided by an influx of data from sensor-based approaches (from Affectiva and other MIT startups). We cannot expect a quick five-hour web course in team skills to be enough, the same way that we cannot expect a five-hour AI course to be enough. Soft skills are not easy to acquire.

Changing the learning experience

How do you become a polymath or foster it in others (say, your kids, your peers, or your employees)? You have to start with changing the way you learn and the context and rationale for learning. First, you have to introduce the importance of developing a personal mission and motivation around learning and what each person wants to accomplish. Then, you need to make it simple enough to make progress so that the learner sees how they are advancing. Third, you need to introduce sufficient incentives that matter to the learner.

Recently, there has been much enthusiasm surrounding the engagement potential of techniques like gamification, online competitions, and learner quizzes, to the point where one almost forgets that these techniques have always been part of the human learning repertoire even if the tools were more limited in scope. I mention these approaches because given the pressure to become Pi-shaped persons, we might need a different educational boost to sustain us. Clearly, four-year degree programs cannot be the (only) answer, although there will be a time and place for structured, time-consuming, all-in learning, too, since some things are and will be complicated to digest in an instant.

Contemporary e-learning platforms—the playing field

Interview Kickstart (https://www.interviewkickstart.com/) is an interesting provider of interview training for new professionals and exploits the fact that interviews don't have an established training protocol in the institutional learning system. E-learning platforms such as Udemy, Skillshare, LinkedIn Learning, Coursera, and Treehouse have mostly adopted the screen-tethered model. Corporate training and education providers (so-called corporate LMS) include top universities, McKinsey Academy, Lessonly, Loop, Gnosis Connect, SAP Litmos, TalentLMS, Articulate 360, Saba Cloud, Adobe Captivate, Cornerstone OnDemand, Docebo, and Learn Upon, to mention some contenders (Pappas, 2015; Trustradius, 2020).

Microlearning—to kick off the process

One highly specific and useful approach is microlearning, the idea that one should train in one particular notion for only a few minutes. Microlearning skyrocketed with the launch of the iPhone in 2007 with apps. Popular learning apps include Duolingo, Khan Academy and YouTube. The social media service Twitter, with its 280 characters and even more with its previously established 140-character limit, epitomized micro-influencing messaging throughout the past decade. Some of us never took to it, but that didn't stop the medium from entering political elections and key societal debates. The Quora concept of ranking crowdsourced answers from global experts to highly contextualized questions is another example. There are strong ties to visual learning in the new microlearning traditions.

Microlearning principles of a single objective per lesson, testing the learner, chunking the content, brevity (3–4-sentence descriptions), short segments (often 3–5 minutes long or shorter), and bulleted format change the subject of learning.

Microlearning output format is typically visual in nature—such as infographics, video snippets and animations, infographics, schemes, illustrations, figures, diagrams—or might even include text messages as the delivery vehicle. Other implementations also exist, including as contextualized QR codes in physical space (museums, classrooms, exhibits). There is typically an attempt to include an "emotional hook" which illustrates how this school borrows from behavioral psychology blended with contemporary design principles. One slight hitch is that a single company, Grovo Learning, has owned the trademark since 2017, which reduces the usefulness and widespread usage of the term itself (though not the practice). A TED Talk or a segment of a structured e-learning course or even a performance support capsule are other examples. In fact, microlearning can be delivered in diverse ways—desktop, video, mobile. It is used in the corporate learning space, but is equally applicable for individuals who watch How To videos on YouTube (Elearning Learning, 2020; Valamis, 2020).

Choosing your distinct Pi-shape

As I have pointed out, becoming a T-shaped expert is possible to learn, but not enough. You should be an expert in two or more domains and deeply immersed in a dozen. Those two expert domains should be relevant to your current main occupation (or should relate to the very next step in your career). One of them should be a technology topic. The other should be a "soft" topic (such as empathy, team skills, language skills, networking). As for what those dozen areas should be, a starting point might be the top 10 emerging technologies that you can identify as marginally relevant to your interests, as validated by market research firms, recently published futurist reports, and your independent research.

Building Pi-shaped groups to formulate projects and make decisions is not just the responsibility of the organization but the responsibility of individuals.

The value of multifaceted interests throughout the history of ideas

Transdisciplinary thinking at the highest level has always characterized the most creative and prolific thinkers in our society. It is instructive to mention a few of them to illustrate how far back into history this goes, and how deeply these few individuals have shaped our world. It was Einstein who said "the greatest scientists are artists as well."

Let's consider a few famous polymaths (Greek: πολυμαθής, polymathēs, "having learned much") throughout the history of ideas (Chiappone, 2020; Tank, 2019, 2020). Several of them were women operating against historical odds—which is no small testament to the fact that it is often contrarian thinking and boldness as a psychological characteristic that makes for the most creative contributions.

Aristotle, Hildegard of Bingen, Ibn Khaldun, Leonardo da Vinci, Madame de Staël, Marie Curie, Ludwig Wittgenstein, Nikola Tesla, Gregory Bateson, Francis Crick, and Richard Feynman would be my suggestions as the most relevant people throughout history to study intensely, in terms of how they developed into polymaths. I've provided specifics about each of them in the appendix but here will focus on a few contemporary examples.

Contemporary period

Mae Carol Jemison (1956–), @maejemison, http://www.drmae.com/, is an American engineer, physician, and former NASA astronaut who also trained as a dancer as well as a social scientist, founded two technology companies, and serves on the board of directors for Kimberly-Clark, Scholastic, and Valspar (Jemison, 2020; Priester, 2020). As a kid, she spent a considerable amount of time in her school library (Biography.com, 2020). Jemison's current project is to make interstellar travel a true possibility within the next 100 years.

Elon Musk (1971–), who is exceptionally literate across physics, engineering, programming, design, manufacturing, and business, is a serial entrepreneur and pathbreaking inventor in fields as vastly different from each other (at the surface level) as e-commerce, automotive engineering, space flight, and energy technology. He attributes his clarity of thought to always going back to "first principles." Starting from his early teenage years, Musk would read through two books per day in various disciplines. He also says you can learn a lot from talking to people, although talking is slower than reading (Musk is big on bandwidth). However, he has stressed how important it is to have a conceptual framework, a "semantic tree" as he calls it, to hang the knowledge on (Stillman, 2017).

Steve Jobs (1955–2011) displayed remarkable persistence. Having started Apple with a high-school friend in a Silicon Valley garage in 1976, he was forced out a decade later. He followed his inner voice so much that he was viewed negatively by many of his employees (and fired those that didn't agree with him or his esthetic choices). Some of his products failed throughout his career, not just in the beginning (notably the Apple Lisa, Macintosh TV, the Apple III, and the Power Mac G4 Cube).

Jobs famously said this at the Stanford 2005 commencement address: "I didn't see it then, but it turned out that getting fired from Apple was the best thing that could have ever happened to me. The heaviness of being successful was replaced by the lightness of being a beginner again, less sure about everything. It freed me to enter one of the most creative periods of my life." He also said: "Don't be trapped by dogma—which is living with the results of other people's thinking. Don't let the noise of others' opinions drown out your own inner voice. And most important, have the courage to follow your heart and intuition" (Stanford, 2005).

However, he did not let failure define him and persisted by returning in 1997 to rescue the company. Jobs combined design, hardware, and software and founded Apple, one of the most iconic companies of our time. Before the iPod, iPhone, and iPad, there were other music-playing devices, smartphones, and tablets, but after Steve Jobs' touch, the whole world wanted to buy these devices. In 2010, during the great recession, he launched the iPad.

Bill Gates had the following learning strategy in college: not attending classes he was signed up for and instead going to classes he was not signed up for. He is famous for being a Harvard dropout. The one thing he has always done is jotting down notes. He uses a version of the Cornell Notes (consisting of three sections: notes with facts, cues with questions, and keywords and summary) devised in the 1940s by Walter Pauk, an education professor at Cornell University (Cornell, 2020). In the system, reciting and rehearsing these notes (using the cues) should happen instantly and should continue from that point. Note taking, of course, also sends the message that you are actively listening (and is the opposite of what taking notes on a cell phone signals, unfortunately, as I have experienced when trying that out).

Today, Gates is the combination of a software geek and now a humanitarian who has reformed the way nonprofits measure impact. He is also about to reform the field of public health, through the Gates Foundation, as well as being a significant startup investor with a notable deep tech portfolio.

Julie Taymor (1952–) is a director, actor, set and costume designer, and puppeteer best known for adapting *The Lion King* into a wildly successful Broadway musical for which she was the first woman to win a Tony award (Britannica, 2020a). Her creative process starts with an ideograph, a

Japanese brush painting, where with "three strokes and you get the whole bamboo forest," she says. The concept of *The Lion King*, for example, she reduces into one image: the circle (May, 2013).

Rowan Williams (1950–), former Archbishop of Canterbury (2002–2012), is a theologian, philosopher, educational theorist, author of four books, including books on theology, and essays, a poet, and speaks 12 languages.

There is no doubt in my mind that had some of these polymaths lived in different times, they would have adapted to the technological vehicles they had available. Leonardo would have founded a series of startups. Ibn Khaldun would have worked with the European Space Agency. Madame de Staël would have started a pathbreaking, elite social media site combining online and offline meetups. Marie Curie would be a life science tycoon.

How long does it take to become an expert?

Since Malcolm Gladwell's book *Outliers* popularized the concept, which was based on psychologist Ericsson et al.'s research on athletes and musicians (1993), many now believe that to become world class in a skill, they must complete 10,000 hours of deliberate practice (5 years full time or, more realistically, 10 years half-time if you have it as a hobby) in order to beat the competition, going as deep as possible into one field. While that may be true for athletes or chess players with talent, who got an early start and work with top coaches, with parental support, and have strong motivation, these conditions may not be possible to reproduce for the average person. However, that shouldn't discourage the rest of us from developing strong expertise in two or three other subjects alongside those we can afford to invest 10,000 hours in. As Ericsson writes, the commitment to deliberate practice is what makes the difference.

It should be said that most people who devote 10,000 hours to anything do not become anywhere as remarkable in that subject as Gladwell or Ericsson insinuate. It is quite possible to play golf at the driving range every night for 10 years and not improve if you train the wrong way or in fact just hit balls without any learning involved. We should also note that most people spend at least 12 years in school and many then do 4 or 6 years of college without becoming much more than average in any field of study.

I happen to think individual variation will be massive—a motivated person learning efficiently could conceivably learn something meaningful within days, learn a basic skill within weeks, start to master a small domain within months, and learn a craft in a few years (the way practical certificates are issued in many professions). A PhD typically takes 4–7 years and I am not sure how much quicker it can be done. I did it in three years, but that was perhaps too quickly to fully digest the complexity I was attempting to cover. Beyond PhD-level expertise, individual differences are so large that there is no point in creating a timescale.

The above are rough estimates; you could put in your own ability and add or subtract 20–80 percent of time depending on where you think you fit. It will also depend on the subject. If you try to become an expert in an adjacent area, it will perhaps match those expectations, but if it's a completely new area it may take longer. If you are exceptionally interested in an area, these timescales may not make any sense, because you might be working near 100 percent of your time on it, just because you obsess over it. Either way, it makes sense to focus and not attempt to become a polymath in one month.

How to make diversity your hidden T

Diversity refers to many things and rightfully so. How does diversity related to the discussion of T-shaped expertise, you ask? Because the whole value proposition of T-shaped persons in the workplace is not related to individuals but to the cognitive diversity and productivity of the organizational unit as a whole (Maitland and Steele, 2020). Cognitive diversity can be a goal in itself but is instrumental in avoiding its cousin, cognitive bias, which leads to detrimental decisions. By discovering and embracing your own diversity (whatever it may be), you have taken the first step. This is not as simple as to say "I am an African-American, or I am a woman, or I am LGBTQ, or I'm a foreigner, so I'm a valuable diversity contribution."

The token value of diversity is just what matters on the surface, and it matters in terms of not discriminating against any one group, which is a separate (and important) issue, but not the issue I'm concerned with in this segment. In terms of innovation, however, what the firm needs is somebody who consistently provides a diverse, different, creative, devil's advocate perhaps, perspective on the challenges at hand. In order to truly deliver on your own diverse identity, which of course is a simplification anyway, after all, belonging to one or more identity categories does not determine your thinking (which kind of is the point with anti-discrimination policies in the first place, let's not forget). Embracing lateral thinking requires effort.

Lateral thinking (De Bono, 1967) usually stems from having had different experiences, having deeply learned something distinct, and having reflected on that learning and then being able to apply it in a contextually appropriate way. Being an effective T-shaped contribution to an organization also necessitates being in the right place at the right time. For that to work, you have to develop a track record of having something to say, so that you get invited to the circumstances where you will be allowed to shine. You need not only stand out in private, you need to stand out in public. You need to be assertive, take risks, take positions, and be prepared to defend them. There's an additional level, wisdom, which is not included. The science of wisdom (Jeste, 2020; Sternberg, 2020) is in its infancy, but whatever elements may go into such a thing, it would draw on implicit knowledge drawn from experience in addition to specific knowledge that you can study for.

T-shaped organizations

What characterizes T-shaped organizations (Wladawsky-Berger, 2015) is obviously T-shaped managers (Hansen et al., 2001) who can recognize talent and plan with that in mind, and who can balance operational demands with collaborative approaches to fix complex, fast-changing problems that also keep showing up.

Figure 7.1 Polymath disruptive model

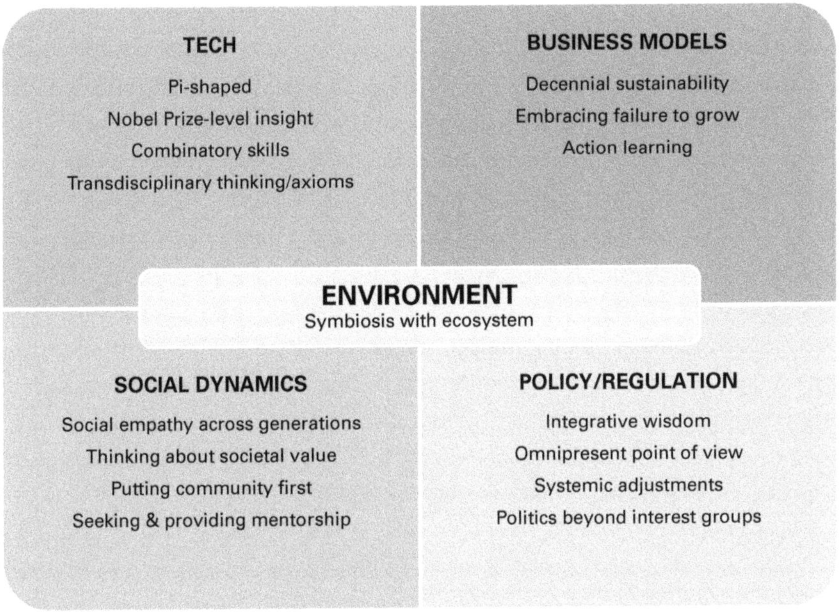

TECH

Pi-shaped
Nobel Prize-level insight
Combinatory skills
Transdisciplinary thinking/axioms

BUSINESS MODELS

Decennial sustainability
Embracing failure to grow
Action learning

ENVIRONMENT
Symbiosis with ecosystem

SOCIAL DYNAMICS

Social empathy across generations
Thinking about societal value
Putting community first
Seeking & providing mentorship

POLICY/REGULATION

Integrative wisdom
Omnipresent point of view
Systemic adjustments
Politics beyond interest groups

Well-functioning teams can compensate for the lack of full-fledged T-shaped persons within the team and so can T-shaped organizations. However, that requires immense collaborative effort and sharing mechanisms that fight information silos, fiefdoms, and ego among domain gurus or corporate leadership. However, that type of rhetoric is often overblown. There are many ways in which such compensation is only theoretical. If you have a one-track mind, the technology you develop will not be broadly interesting or appealing to a wider audience. For that reason, there is no way around gunning for T-shaped experts everywhere in the organization. In fact, without the T mindset, these folks may not even be working on the right problem. Schools such as MIT realized this long ago and hire professors who are adept at both mind and hand, theory and practice, and research and commercialization. That is a much faster path than relying on murky organizational communication patterns anyway (Wladawsky-Berger, 2015).

Conclusion

T-shape matters more than ever, but not the simplified version you read in blog posts, and not for the simplistic reason of "getting ahead" under which rubric this type of argument is typically presented. The deep and wide Ts are the most valuable, the true Pi-shaped person is what the next iteration of society will be built on, technology or not.

Society especially needs more polymaths in tech, both to continue the rapid innovation cycle we are in and, more importantly, to capture broader parts of the population with tech innovation that responds to people's true needs, not just what sells in the short term. Moreover, society needs more understanding of the interplay of technology and society, which—in itself—is a T-shaped discussion because it involves a minimum of two separate disciplines and experiences.

In this chapter I looked at how the T-shape has mattered in the history of ideas of art, science, and technology. While using the nice conceptual clarity of the T-shape concept, I argued that the T-shape is not enough. Rather, you need to evolve into a polymath. I have some very concrete examples of contemporary polymaths and some hints about how I think they became such. The task is now yours: go do that. However, I'll be nicer than that. There are a few more chapters left.

In the next chapter, I shall look more closely at how you can personalize your insight ecosystem to implement your new T-shaped identity, redouble your efforts at becoming a double T and beyond, and set yourself up to continued success in the workplace—and in life itself—by becoming truly transdisciplinary.

Personalize your insight ecosystem

08

In this chapter, I look at how you can personalize your insight ecosystem to implement your new T-shaped identity, as well as your polymath ambitions, and set yourself up for continued success in the workplace—and in life itself. First, I define insight. I then discuss the traditional tools available to learn technology industry information and developments, including market research, search engines, newsletters, events, consultants, expert networks, partnerships, and foresight. Following that, I specify what I see as the key elements of a more personalized insight toolkit which is tailored to yourself because you have to be in balance first, secondarily to the challenges you want to solve, and thirdly to the organizations you work with to achieve those goals. Nobody can solve all problems and certainly not at the same time. Therefore, focus is necessary, even as you set the aperture wide to take in a lot of impressions so your response can be creative, novel, and compelling. Toward the end of the chapter I talk about criteria for establishing (and reestablishing) focus. Finally, I discuss how technologists humanize the discussion, enabling us to see the motivation behind each discovery. In the appendix, I give recommendations for who to track.

What is insight?

Insight is, in itself, a problematic concept. What does it mean to have insight? Does it mean to know something better than those around you? If so, how do you explain it to a wider group, even your own team? Acts of persuasion often requires sustained co-presence and momentum as well as numerous props, allies, and fortuitous circumstances and a sudden open window of opportunity. You need to jump through it before the window closes.

Many consultants I know, and an increasing number of corporate technologists and strategists, subscribe to publications that cover specific topics related to their job: world newspapers, business publications, industry blogs, news about their current and past clients along with their competitors.

Readily available insight tools

There are many tools available to build insights about the competitive environment and to track trends and emerging developments, including (a) typing in queries in consumer search engines, (b) subscribing to free newsletters and market research, (c) relying on trade association newsletters and gatherings, (d) attending industry events, (e) hiring strategy consultants, using on-demand expert networks, (f) relying on strategy, internal consulting, or R&D teams, (g) seeking open innovation partnerships with universities, accelerators, VC firms, etc., (h) seeking guidance on foresight, forecasting, or scenario planning, (i) taking online courses, (j) listening to podcasts, (k) taking on a mentor or tutor either face to face or online, and (l) maintaining high-end subscriptions to customized databases and custom trend reports.

Each method has advantages and disadvantages and you cannot do all of them for all topics you are tracking. There is no relationship between cost and value. For example, using a consumer search engine, listening to podcasts, and tracking online influencers is often free but can yield great insight if you take the time to select well. Meanwhile, personalized mentors and tailored on-site strategy consulting yield lots of insight, too, but are quite expensive.

In reality, given today's shifting knowledge needs, none of these works in isolation. The market for online content discovery tools will continue to grow as the complexity of navigating online content increases. The only sure thing is that tools evolve, and so should you. My view is that the winners of tomorrow will have access to an e-enabled, personalized growth toolkit, likely based on highly advanced artificial intelligence that fetches information based on a sophisticated set of constraints and requirements that become computer algorithms that operate 24/7. However, it will not replace your own efforts, at least not in the next decade. What does that mean?

Simply put, you need a diversified approach to acquiring and processing information, deliberating decisions, and scaling your implementations.

Whatever you do, don't go it 100-percent alone. Don't try to track everything, learn everything, and summarize everything on your own. Work with teams, peers, networks that have similar interests. Without assistance, you will surely fail, no matter how advanced your insight was to begin with.

Your personal insight and growth toolkit

Oliver Wendell Holmes, Sr. (1804–1894), the American physician, poet, and polymath based in Boston, acclaimed by his peers as one of the best writers of the day, had this to say about insight: "A moment's insight is sometimes worth a life's experience." I think that sets the stakes for what I am trying to get at in this chapter. I am not looking to say that you should amass as much information or knowledge as possible—even in your domain of impact. This is not a question of quantity. I am also not saying that you need to be creative for creativity's sake.

What the game of attaining sufficient insight to approach wisdom is about is to make a uniquely fashioned imprint on the world that represents the very best that you can be, and that at the same time makes a contribution beyond that which is to your benefit alone. That is what wisdom is.

The struggle to stay up to date

Tracking retail used to be straightforward. If you had your own retail operation, there was a constant stream of input feeding in from sales, marketing, operations, and the point of sale. Tracking trends would mean following the main industry associations, attending a few key events every year, and talking to colleagues and competitors informally throughout the year. Trends would shift every year or sometimes seasonally, two or four times a year at the most. Megatrends last for decades.

According to industry analysts and marketing firms, the most salient trends are always tied to a generation. Millennials over the last decade. Post-Millennials if you sell to young people today. That makes it neat and organized. Know which generation is your target customer and you know how to market and sell to them. However, it may have become more complex than that. Generations are molding together. Technology moves faster. Competitors arise every month.

Catherine, the head of a global retailer, had been sweating it for months. Investors and markets were not optimistic on the sector's outlook. For a while, she had positioned the company she runs as an outlier, a growth machine despite the massive changes in shopping behavior and the media stories about closing thousands of retail shopfronts across America. At that moment, it would seem the trends were catching up with them. They had just announced a round of layoffs and store closings and the sale of a specialized retail arm. Numbers were down for the third quarter in a row. Catherine needed to pull something out of the hat. She needed to switch on growth. The realization came during a conference she attended at the MIT Media Lab. She had not planned to go but last minute she found three hours in her schedule and felt like taking a breather from all the taxing issues and just go to the well. They say MIT is like a firehose. The Media Lab has the additional distinction of being where Nicholas Negroponte wrote his bestseller *Being Digital*, the book that started the massive shift toward digitalization across all sectors of industry.

The event was called "The Future of Tech, Analytics and Consumers." Catherine found the topic and speakers enthralling. Particularly, there was one speaker from MIT who truly had seen the light. Her experiments showed that shoppers increasingly wanted to co-create the products they ended up buying and promoting. Fascinating. She was reminded of one of the corollaries of sales: You sell more if you truly embrace the mindset that you are helping people you are not selling. People need and want help and might even appreciate it. The reward for helping is to spend. In the end, everyone wants to feel good about themselves. With digitalization, customer input can become scalable, truly transformational for product development.

But users and customers also want to be involved in the creative process. They want to be product developers themselves. They want to have impact. What is better than feeling that you are having an impact and working with a global retail brand? Catherine saw the potential, but didn't quite see how she could get there, given her balance sheet obstacles of the moment. She needed more than a three-hour breather. She needed help herself.

Here's what Catherine did. She turned to her organization for answers and she openly solicited outside assistance, quietly, but not invisibly. She tracked the retail industry trends both from internal data sources and from external trade associations, research firms, and academia.

How to cultivate your own insights

Few people are both thinkers and makers of companies, products, and high-value services on a global scale. Those who are can usually be found in history books, memoirs, and biographies. Elon Musk, Steve Jobs, and Thomas Edison spring to mind. The surprising fact is this: Everybody else has access to the general information about these heroes of modern science and technology. Also, in many ways, the information each of these polymaths was working from was not in and of itself so different even from what their peers had access to, sometimes even less. For example, how do we explain that the two brothers Wright, from an unremarkable upbringing and context, could outcompete everybody else and be the first to create an airplane? We still don't quite know how it happened.

In order to excel, you either need to dig deeper into what created those minds, or you need to tap into different insight, a few layers down, slightly more invisible to the general public.

To dig into these thinkers, you can read biographies, read their original works, interview people who worked with them, try to copy some of their learning styles and habits. However, it is unlikely that merely reading a book about someone will change your own habits. To make a true difference you would need to commit to changing your life, gradually molding into the basic habits that characterize thinkers and doers of stature. For most of us, that's too much hard work. I'm not advocating that this is desirable and possible for all of us at all times. But what is possible is to make maximum effort where we have the chance. Everyone can set a clear goal of becoming 10 times more insightful in a particular domain and can get there in six months or a year. Especially if you define the milestones at the outset and validate these milestones with somebody who knows what they are talking about, perhaps a friend or colleague who can act as an informal mentor and yardstick.

From insight to growth

Insight for its own sake is something academics may pursue, humanists or literature fans might think this way, but business and technology people usually have a more utilitarian goal in mind. How to capture value? How to create a product? How to win a big contract? How to get a promotion? To achieve any of these goals, you need to grow. They don't happen by themselves.

How does growth work these days? How can you tap into it? The answer many companies turn to is two-fold: tackling the inside-out challenge and tackling the outside-in challenge. The inside-out challenge consists of better understanding your own business, including digging out better analytics from your own data. The outside-in challenge consists in better understand the environment using external sources, partners, startups, and university relationships, focusing on trends, research, or foresight. The combination is a high probability bat for growth through insight.

Seeking new sources of growth is typically new and challenging. It forces companies to rethink received wisdom. It hurts. What one finds may even invalidate the experience of senior leaders within the business. Some do not catch up. Some resist. Either way, it is painful. But some companies do not even get there and are poised for perennial decline. What is the difference between those companies that grow and those that don't? What characterizes those that do not, those that face change head on?

The way to track insight for growth

Having looked at the options available, and the grand ambition of personalizing growth, followed by an example of how hard it can be, I'll now recommend how to work with the most effective insight tools in order to achieve growth.

The role of trade associations in tracking insight

The National Retail Federation (NRF), the world's largest retail trade association, might be one starting point. Its members include department stores, specialty, discount, catalog, internet, and independent retailers, chain restaurants, and grocery stores. The NRF's resources include regular events and White Papers intended for members or for lobbying. However, the NRF's perspective is obviously slanted toward making the sector look good and one cannot expect to find perspectives that go against industry interests there, and certainly not negative news such as opinions on which retail subsectors might have the worst growth prospects. The association clearly want to keep its members. It is also only one side of the story from one source.

Strategy consulting insight

Many research and consulting firms have retail divisions, and some are retail only. Judging from the top hits on Google, one might think Deloitte, KPMG, or PwC are leaders in the space, as each issues its annual retail trends publications. However, be watchful. There seem to be four kinds of search results that may turn up.

The first is typically simply a marketing page tooting the horn of their expertise in the area, which is only marginally helpful and would mean you need to search around more or pay up for an actual project.

The second is a promotional White Paper, typically between two and five pages, which usually does have some insight and even some figures. If you look into it, it is also just a shopfront and typically names individual experts at the end of the White Paper for further elaboration or leads you to a download page where you may need to buy the full report.

The third search result leads to an abstract or summary of a report which you then are asked to buy a license to read. There are two such pages, vendor pages such as Gartner or Forrester, or third-party providers such as Research and Markets that give access to reports from many vendors but where you may not know which vender you are buying from until you have paid for the report.

The fourth result is the home run, the full-text report laid out in 30–100+ pages where somebody deploys a content marketing approach and has invested in some amount of work for a high-quality yet free product.

Many of the best reports of this kind are annual. These vendors are thoughtful about building a reputation as a knowledge provider over time.

McKinsey, the top strategy consultant, may be a starting point. Hiring its experts is expensive. You may be able to do it once every few years, if you are lucky. However, you most likely cannot use them on a regular basis, unless you belong to the top echelon in a Fortune 500. Having said that, the good news is that you don't have to purchase their services to take inspiration from McKinsey.

Even McKinsey needs to market its services. This means it gives away a lot of information for free. You could choose to browse its free content on the subject of retail online. In this case, the online site does have quite a bit of content, so that is promising. Very rapidly, however, you reach a paywall. Your only option then is to contact the company for a million-dollar project looking into the sector for you. That's why you need an alternative approach.

It would be tempting to go a few tiers down and simply hire cheaper consultants. The challenge is that with consulting you tend to get what you pay for. An expensive consultant is usually a good consultant. A cheap consultant is not necessarily a bad consultant, but definitely somebody who is less networked, who may be outside the main currents where information is exchanged. Some of that information may be crucial. At least, that's how it used to be. But things are changing. Lots of smart people don't want the McKinsey lifestyle. Perhaps they worked there for a year and are now with a lower tier, or they are independent.

Asking an expert you know, a friend, an acquaintance, for some free advice is another favored approach. As it used to be, you had to call the consultant you had on retainer, or lacking that, call the local university and ask a professor what he thought about the issue. Failing that, of course, make a call to your own network of peers, connections, alums, friends, or colleagues, anybody you knew that might have an inkling and who might have a moment to spare to discuss.

Where can open innovation contribute?

Over the last decade, a flurry of open innovation platforms has emerged, places like GLG, Catalant, or a myriad of other expert networks where you can contact an expert on demand who will answer your question for $100, take a call for $500, or take on a project for $1,000 or $10,000, depending. Common with most of these approaches is that you tap into a wider net of experts, pay a bounty, and get advice from the experts in the crowds surrounding an issue.

Using search engines effectively and wisely

What if you looked for the highest-quality online content in retail instead? You might imagine doing a Google search for Retail Trends, for example. Without any further qualification, the 244,000,000 results (at the time of writing) would mean you would spend the next few hours sifting through pages and pages even among the first few hundred results, aimlessly searching for keywords and brands or other known indicators of relevance and quality. The first few hits would be ads, i.e. proceed carefully and only if you know the brand. The next few hits would be sites that are search engine optimized (SEO) and typically either large brands with huge resources or small niche publications that have gamed Google with artificially elevated inbound links relevant to your search topic. Obviously, somebody with great search engine user skills would be able to further dissect results based on an optimal combo of keywords, i.e. retail AND trends AND technology AND pdf, for instance, but this is cumbersome and fraught with the risk of missing important documents.

Regardless which kind of publication or offering you are looking at and intend to let influence your next step, you had better assess the motivations of the source. Industry positioning strategies of vendors in this area include visibility, thought leadership, or also more activist strategies such as wanting to shape the industry in some desired direction.

The connection between tapping into insight and switching on growth is often assumed without any evidence, which is a mistake. One cannot simply take for granted that seeking insight leads to insight or that having insight itself leads to the kinds of actions that generate growth. For one, there is the challenge of execution. This is where most expertise-based approaches typically fail and have failed in the past.

This is how it goes. Reasonably good advice is given, often at a hefty price from a top-tier strategy consultant. Having listened to a presentation about what to do and not remembering much from such a presentation, you are left with the slide deck, with recommendations. Some time goes by and you implement absolutely nothing. Not because you do not want to or do not understand what to do. You are not alone in the decision, perhaps, and time goes by and other more pressing challenges push the current one lower on the priority list. You may think you understand what is meant by a recommendation as well, and then when you implement, you may or may not realize that you are not at all working on the right problem.

The context of the challenge may have shifted ever so slightly or simply the process of interpreting the solution to the challenge means you are doing something else other than originally intended.

Syndicated information

For some time, the way to go would seem to be a technology called rich site summary (often called really simple syndication), now commonly known simply by its abbreviation, RSS, or indeed a "web feed" if you prefer the food analogy. The point is when online content is set up or delivered in this particular way, it allows users to access updates to online content in a standardized, computer-readable format. These feeds can, for example, allow a user to keep track of many different websites in a single news aggregator. The whole online content marketing industry is built on this foundation, which makes it easier to appear as though you are an important player in the news or knowledge ecosystem, as an intermediary who chooses what to present as the most relevant at any given time, or indeed to become a news agency, a "Reuters" in their own field. The popularity of RSS has had its ups and downs. Nowadays, most users do not have to even think of RSS because news is fed to them through the platforms of which they are part. LinkedIn News, for example builds on a combination of RSS and user-generated content. One medium, however, is 100-percent reliant on RSS and that's podcasts.

Podcasts

Over the past few years, RSS feeds have seen a renaissance because of podcasts, where dozens of podcast players (Apple Podcasts, Spotify, Pandora, and iHeart Radio being the main ones) often are willing to syndicate your podcast and offer your content across their platforms, all using your RSS feed which goes back to the one central location where you originally published your content.

Podcasts are indeed an emerging source of insight because of the intimate way (through voice) the content is delivered as well as the context of delivery (you could be running, just getting out of bed, putting on makeup, or using it right before you fall asleep), which are all highly precarious parts of your day where you are particularly prone to consider what you are being presented. I give a shortlist of some particularly insightful podcasts in the appendix.

Online courses

The idea of remote learning has had many form factors throughout the past century, from correspondence learning over what we now call snail mail, through radio, TV, cassettes, DVDs, and the internet, which is today's preferred format. I know people who have taken hundreds of online courses. They are often inexpensive (even free) and tend to give high-quality instruction which you can receive without moving out of your chair. You can even be in a class setting and get feedback from an instructor or be graded and perhaps get a certificate. There is sufficient motivation for millions of people to engage in this type of medium. Today, you can find a great variety of courses. Taking a course could be a way to get into a new domain, a way to refresh knowledge, or a way to supplement other types of learning. Either way, it is a fantastic resource which greatly enhances your ability to stay up to date on the future of technology and business. But you have to pick the right courses and you still have to do the work. Learning does not come automatically.

Technology fads

On a final note, technology fads are common. They are created by virtue of the fact that when media and investors jointly get excited about a space, they bring along others for the ride who may or may not be fully aware of the underlying logic at play. Investors, in turn, become disappointed by the return on investment. Media, by its very own logic, which is somewhat deterministic, needs to move on to a new topic even if it is still, deep down, enamored by the second latest, most favored topic.

Time and again, this has happened to Nanotech, Cleantech, Edtech, Medtech, and AI. In fact, probably it has, in some way, happened to all of the technologies we are discussing. This is the reason I truly feel Gartner was so apt to coin the idea of the hype cycle. At least now we are all somewhat aware that we are part of a hype cycle. But that is not the same as having confidence that we can always tell where a technology currently stands in a given hype cycle, or even in the concept that hype follows anything close to Gartner's cycle logic.

Three approaches to increase your sci-tech literacy and awareness

In the rest of this chapter, I outline three approaches to better understand the disruptive forces of science and technology. As part of that effort, I recommend a deep dive into taxonomies—the way technology knowledge is conceptually organized. Being familiar with the most recent taxonomies is a time-saver when doing research and will also begin the process of making you literate enough to speak to a tech expert in an emerging field. Having the right set of questions is the absolute most important challenge in tech disruption. If you have good questions you are not only able to detect trends but may begin to foresee them and eventually take part in creating them.

Tech discovery approach #1: personalize your approach

Whether we are aware of it or not, each of us has our own, personalized *disruption matrix*, reflecting our current concerns. It may not be as system-atic as the one in Figure 8.1 (which is just an example), but it should at least be at that level of granularity. However, if somebody claims to be able to keep more factors in focus at any given moment, they are likely deluding themselves.

Figure 8.1 Personalized disruption matrix

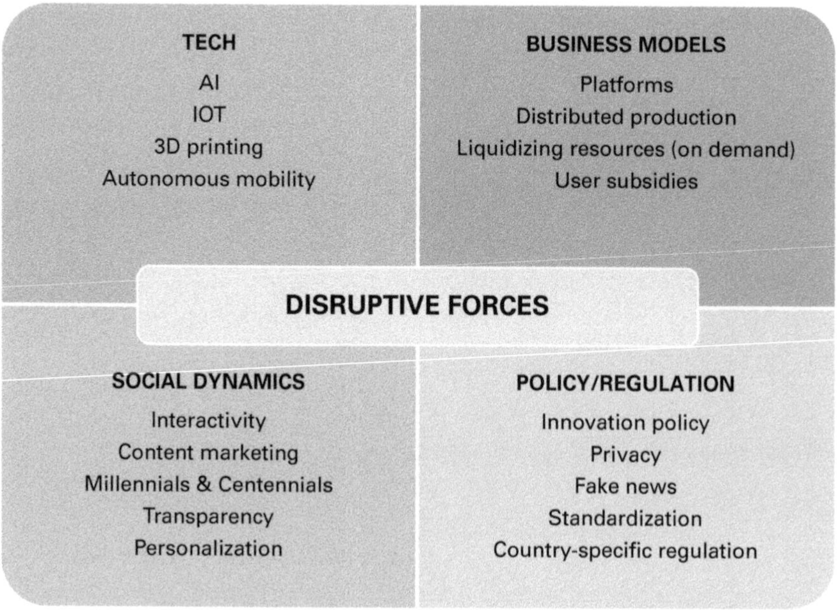

TECH	BUSINESS MODELS
AI	Platforms
IOT	Distributed production
3D printing	Liquidizing resources (on demand)
Autonomous mobility	User subsidies

DISRUPTIVE FORCES

SOCIAL DYNAMICS	POLICY/REGULATION
Interactivity	Innovation policy
Content marketing	Privacy
Millennials & Centennials	Fake news
Transparency	Standardization
Personalization	Country-specific regulation

My recommendation would be to create a minimum of four such disruptive forces maps, one general one that has everything you are tracking in one map, and one for each technology you are responsible for tracking or have the ambition to learn more about in the next few months. You should try to validate the set of subforces with an expert or two, either a colleague, a friend, or a mentor. That now becomes your *mental map*, your mental model not only for how you see the future of that technology but for how you go about seeking further insight.

The challenge is that these "mental maps" are rarely shared, so even close team members are discussing without in-depth awareness of each other's true concerns and situational understanding. Spelling this out is a fruitful team exercise that I use in my innovation workshops. Seeing how this clarifies people's projects and starts to shape a higher level of cognitive cohesiveness at a group level is astounding.

You can also experiment with putting up your personalized disruption roadmap on a wall to keep jotting down observations you have along the way. Drawing relationships between concepts has proven to be a fundamental way to remember things but also to deepen your thinking beyond that. Attempting to teach your current understanding to others is the next level. By doing so, you will start to question your own assumptions, clarify your points, and begin the next level of inquiry. It is surprising to see that these methods are still not institutionalized either in business or in education, given the increasing evidence of how visualization is a key part of developing deeper insight.

Tech discovery approach #2: develop your insights beyond 101

In the "age of the internet" some subscribe to the notion that all schools should do is "teach you how to learn" because the internet somehow will have all the actual empirical answers you need to get through your life, even in highly complicated fields.

Unfortunately, nothing could be further from the truth. Without a prepared mind, which both contains basic science and technology classifications and a host of facts about the relationships between those classifications, facts, and insights, browsing the internet, even if you had the right search terms, wouldn't teach you much. For instance, without knowing how nano science or quantum computing categorize current scientific challenges, how can you invest in them? You would be forced to rely on second-hand interpretations, which introduces possible delays and errors. Or you could be investing based on indicators that don't depend on your insight into what the approaches are actually doing, which is a risky proposal (but not an unprecedented one).

However, there are many ways to prepare your mind. There doesn't have to be only one gateway. You might find one approach useful for one problem and an entirely different one equally useful for another. The situation, even your mood, might dictate how you are most likely to learn. Maintaining a certain flexibility is instrumental for grasping science and technology in action. Even the most mundane observation might unlock a deep truth. The reason is that all innovation advances through imperfect prototypes that need to be further refined. It might even be that the final state is not the most interesting. I'll come back to the importance of personalizing your approach to generating insight in Chapter 9.

Tech discovery approach #3: understand the scientist as well as the science

When studying science and technology, it has for some time been uncommon to focus on the person behind the discovery, even though this was how one learned these subjects a century ago, and quite efficiently at that. Perhaps because of the emphasis on the scientific method as well as upon the most up-to-date findings, most of us learn science as the history of ideas and discoveries. Even worse, we sometimes just memorize techniques to which we know the answers from the back of the book. There is no learning involved in that. It merely serves as a time suck, a crossword puzzle you don't enjoy. It will not help you understand how science or technology might evolve.

In the process, the actual way science developed gets obscured. In textbooks, you are led to believe it was a linear process: *a* builds on *b* which leads to *c*, instead of a confuses the relationship between *b* and *c* which are found to be faulty and lead to *d* and *e*, which later is all falsified by *f*, although *f* is not known to scientists in X lab in Y country, so they continue along the path of *d* and *e*, which later is found to be fruitful for certain limited cases. Such was the case with most discoveries in both science and technology, from major physics paradigms to experimental chemistry.

Furthermore, as Firestein (2016) has shown, all of science is based on a series of failures, called *experiments*. There is enormous integrity in failure, Firestein writes, and it enables science to get wiser the next time (if a negative result is communicated, which is somewhat disincentivized in modern science and that is a problem).

However, I believe it is uniquely meaningful and instructive to read the writings of the scientists themselves. The reason is simple: It helps me understand what they were worried about, what motivated their thirst for knowledge, and what, ultimately, quenched their thirst. They spent their time wrestling with these problems because these problems matter to them. And understanding what drove them gives me an appreciation for how they progressed and inspires me to learn deeply, passionately, and knowingly on my own. Today, the easiest way to grasp this logic is through reading biographies and autobiographies of leading scientists.

New ground is continually being broken by notable scientists in the areas of genomics, nano science, computer science, and economics. Understanding what problems these notable scientists feel are the most critical to consider, and being acutely aware of what they have yet failed to understand, can help you better understand—even as a non-expert—where the field is going. So, if you are looking to track trends and innovation in the area of science and technology, it is critical to keep an eye on the scientists behind those developments with an eye toward gaining insight into what problems they are addressing.

There is even the argument, and Firestein (2016) makes precisely that argument, that by going back into old papers and looking at where the scientists describe their failures, you have a treasure trove of promising experiments to replicate 10 years later, perhaps with improved experimental setups, knowledge, and technology, perhaps with pathbreaking results.

Compiling your list of scientists to track

Now that you understand why it's important to track the scientists, how do you figure out who to track?

Over the past year, I decided to try to identify the top 1,000 emerging minds of science and technology with the idea that I would start tracking what they were up to. I cast the net far and wide in terms of backgrounds and was highly selective in terms of proven achievements, focusing on highly vetted disruption experts across a wide variety of fields, such as influencers, industry experts, entrepreneurs, academics, technologists, scientists, venture capitalists (VCs), authors, and speakers.

The biggest challenge was finding those who would stand out. I identified lists of the top people in individual science and technology fields, as well as those who had won prestigious prize competitions. You need a bit of network knowledge to figure out which of these lists are reputable or you need to study citation indices in each field, which is quite a bit of work. Being comprehensive with this is not a one-man job, which is why I used man/machine power, including crowdsourced assistance. I also reviewed results of prize competitions that I believe are signals for potential extreme disruption potential, with the top 20 appearing in the list directly below. Once I identified such extreme disrupters, I scored them on insight, education, experience, authorship, expertise depth, knowledge breadth, creativity, and public speaking, and normalized the total score on a 100-point scale. I've also created a software platform to track these people.

Top 20 indicators of extreme tech insight network disrupters

CB Insights unicorn startup founder

Committee 100

Forbes: The World's Top 50 Women in Tech

Forbes 30 Under 30 (Science)

Knight Science Journalism Fellow MIT

Lemelson-MIT Prize Winner

Nature's 10

Nobel Prize (Mathematics)

MacArthur Fellows

McKinsey Practice Leader

MIT Tech Review Innovators Under 35

Member of the President's Council of Advisors on Science and Technology

Stack Overflow's Most Influential in Tech

TED Fellows

The Midas List: Top 100 Venture Capitalists

Webometrics.info Nanotechnology Experts

WIRED Global 1000

White House Fellows

WSB Future Forward Thinkers

Vanity Fair Future Innovators

WIRED 2017 Smart List

As you can see, the list spans the public and private sectors and nonprofits. However, the biggest surprise was that in setting the bar very high, I found there weren't that many prize competitions that made the cut. I quickly discovered that many competitions don't publish their criteria, are more concerned with picking already famous people, or simply don't seem to be predictive of future success. Counterintuitively, Nobel Prizes are the least helpful to me because the significance of the Nobel Prize lies in past achievements. What I'm really after is the people who haven't peaked yet and have worked with a former or existing Nobel Prize winner in Physics, Chemistry, Economics, and Medicine. There's credible research proving that those who have worked in the lab of somebody who went on to win the Nobel Prize are more likely than any others to also themselves win the prestigious prize. Those experts might be harder to identify, but they are also more exciting, innovative, and futuristic.

Once I identified such extreme disrupters, I scored them on Insight, Education, Experience, Authorship, Expertise Depth, Knowledge Breadth, Creativity, and Public Speaking, and normalized the total score on a 100-point scale. I have listed a few of the thinkers I discovered and their fields below.

Extreme disrupters

Eric Paley (97/100) is a notable investor in enterprise IT innovation and
startups.

Vladimir Bulovic (95/100) is a world-leading expert in nanotechnology and a serial founder of startups.

Rana el Kaliouby (92/100) is a unique voice in emotion AI and startup innovation.

Elon Musk (91/100) innovates in both autonomous driving and space flight (and more).

Sangeeta Bhatia (91/100) is a pioneer in biotech entrepreneurship.

Azalia Mirhoseini (91/100) is transforming chip design for faster AI.

Grace Gu (87/100) is leading the race toward bio-inspired materials.

The strategy becomes a unique knowledge acquisition one only when you start strategically and methodically following people who are notable *in fields which you know very little about.* Tracking people in your own field can also be useful, but it is more mundane from the perspective of deriving polymath-relevant transdisciplinary wisdom (see Chapters 7–9 for individual insight strategies).

On my end, I started tracking experts in biotechnology, AI, and a host of other trending sci-tech areas. I was surprised to find that it was possible to follow their progress without specialist knowledge in their fields because these scientists were quite good at simplifying their findings for non-experts. This resonated with what I have found from working with MIT's top professor-innovators, which is that those masters at the top of their game thrive when given the opportunity to tailor their scientific findings to a non-expert audience. We should all be so lucky as to have an advanced sci-tech field we could explain to our grandmother.

Conclusion

Personalizing your insight ecosystem is essential. Without a well-thought-out system to sift through current events, disruptive forces, threats, and opportunities in your field, you are going to be less efficient. By not having a wider lens to adjacent fields you also risk getting disrupted.

The bar to stand out is high. A lot of news and information is available for free to all who care to look for it. Yet, there is arguably information overload, so those who know how to prioritize information are in a better position.

As you ponder how to personalize your approach to the various tools available (such as market research, search engines, newsletters, events, consultants, expert networks, partnerships, and foresight), the important thing is to be purposeful, periodically review whether your approach is working, and track new entrants in the market that might simplify your approach. Finally, the proof is in the pudding: Are you successfully staying ahead of the competition or can you demonstrably show you know the trends and disruptive forces in your chosen fields? More importantly, what products or insights have you produced (this week, this month, this year) by combining knowledge from the two or three disciplines you specialize in? This is where a mentor might come in handy, some independent person who can come in, track your progress, and comment on what they think you should do next, or better, listen to you as you chart out what you want to do next.

Key insights and takeaways

1 If you were to create your own, personalized disruption matrix, reflecting your current concerns, what might it look like? Try to fill in at least three subforces within each of the four boxes—tech, policy, business, and social (see Figure 8.2). What has shaped your understanding? What new info or events are likely to change this mix of factors?

Figure 8.2 Disruption insight matrix exercise

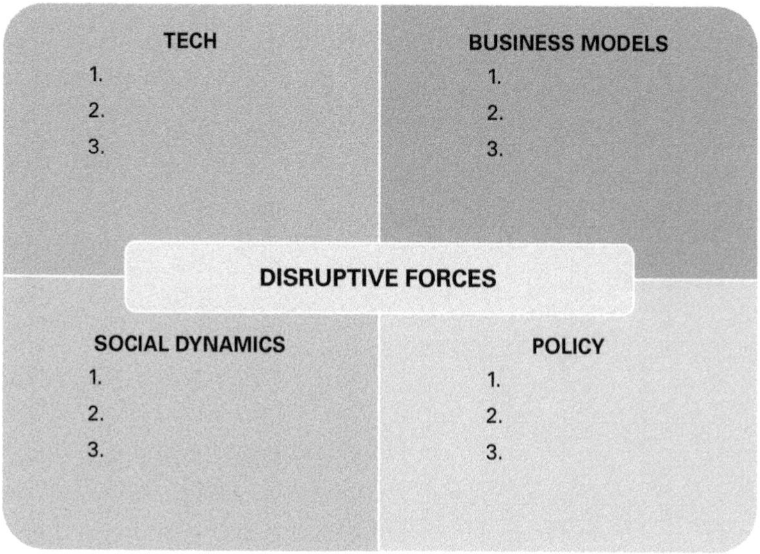

TECH
1.
2.
3.

BUSINESS MODELS
1.
2.
3.

DISRUPTIVE FORCES

SOCIAL DYNAMICS
1.
2.
3.

POLICY
1.
2.
3.

Merge with technology to achieve a cognitive leap

09

In this chapter, I look at how to proactively respond to the impending future of technology to ensure you take a productive role in the next decade.

As I see it, there are five key roles available to you as this decade progresses. They are (1) to develop new machines, (2) to tweak and use machines, (3) to actively *integrate* with explicitly chosen machines in order to achieve cognitive leap, (4) to set *boundaries* for the entry of machines into your life or that of your significant others, and (5) to take on the *stewardship* and ethics of machines. The five technologies I have analyzed closely in this book—AI, blockchain, synthetic biology, robotics, and 3D printing—are currently, in their most advanced forms that is, more prevalent in secret government facilities, the military, research hospitals, university labs, biotech labs, or highly experimental startups at the forefront of neuro-engineering. That is a big challenge—we cannot afford to have such a small group of people concerned with things that might change the course of humanity in this century.

In this decade alone, almost all humans will be exposed to colleagues, friends, and significant others who have one or more of these roles. There is also a possibility that the future of manufacturing or indeed the future of work among white-collar knowledge workers will depend on significant mergers with cyborgs. However, even if these are general directions, you can shape your reactions, and you can choose to help accelerate or decelerate. Each will have a cost.

This decade will be one of continued man/machine symbiosis whereby most of us will gradually and increasingly merge with technology. For some this will be a passive process, for others it will be actively desired, and also, to some extent, encouraged by others around us. In any case, the reflection level on what is happening will vary, and without preparation, objecting to the general technology-intensive nature of our world (if this is what you choose) will be both emotionally costly and unnecessarily complicated.

Integrating with machines

The merging of humans and machines is gradually happening throughout this decade (Prabhakar, 2017). However, we have been so-called cyborgs (e.g. part human and part machine) since the first artificial limbs were successfully attached to the human body, which goes back to the origin of mankind. For our purposes, however, it is the introduction of a conduit of electronic communication between the limb and the body's nervous system that defines a true cyborg.

The oldest neuro-prosthetic experiment on record is from 1957 when otologist C. Eyries operated on a deaf person in Paris to try to improve his hearing. Eyries did not install a cochlear implant per se but rather an electrode in an attempt to stimulate the patient's auditory nerve. Even as that experiment was largely unsuccessful, it spurred further experiments throughout the 1960s. In 1964 a six-electrode implant was tested in California, and clinical trials started in 1970; single-wire implants were trialed in 1974 and in 1978.

By the end of the 1980s cochlear implants had become "the predominant treatment for profound deafness in the United States, Europe, and Australia" (Mudry and Mills, 2013). The approvals by the FDA (in 1984 and in 1990) greatly helped the market take hold, although it is in the elderly market that the device has found its true home, given that many see the hearing impaired (formerly known as deaf) as a linguistic community rather than as a group of patients. Some in that category were not actively seeking treatment the way the early experimenters had assumed (and hoped). This fact is a lesson for any sci-fi-inspired medical or neuroscience innovator—designers have to ensure the social dynamics are favoring the product being made (Institute of Medicine, 1995).

The process is now accelerating with the connection of Internet of Things-connected sensors and digital interactive capability of more recent medical and non-medical wearables and devices.

Toward the end of the century, artificial skin, engineered blood, and fully functional synthetic organs are expected to appear and become part of live experimentation in humans. By that time, we can expect mind-brain implants, such as the ones developed by Elon Musk's startup Neuralink, and others from startups Akili, BrainCo, Emotiv, Kernel, MindMaze, NeuroSky, and Paradromics to become commonplace at least in neurological patient experimental treatments as well as for voluntary use and experimentation by the technology's developers (Choudhury, 2019).

Mind-brain implants and other startup approaches

Boston-based Akili, founded by Adam Gazzaley in 2011, builds clinically validated cognitive therapeutics, assessments, and diagnostics that look and feel like video games.

Lausanne-based MindMaze, founded by Indian-born neuroscientist Tej Tadi, designs an intuitive mind/machine interface through its neuro-inspired computing platform, developing medical-grade virtual reality products to stimulate neural recovery and able to capture neural signals based on intent.

Austin-based Paradromics, founded by Matt Angle in 2015, is the developer of an implantable brain-machine interfacing technology ("modem for the brain") intended to increase data transmission rate between brains and machines for therapeutics. Human clinical trials will start in 2021 or 2022 (Lalorek, 2019).

San Francisco-based NeuroSky, founded in 2004, builds health and wellness biometrics for mobile solutions, particularly through its exercise equipment for children's minds.

Los Angeles-based Kernel is a neuroscience company founded in 2016 that specializes in developing brain-recording technologies delivered as neuroscience as a service (NaaS).

Boston-based BrainCo, founded by Harvard brain science PhD Bicheng Hang, strives to apply brain machine interface (BMI) and neurofeedback training to optimize the potential of the human brain. Its main advertised benefits are studying more effectively, less stress, improved focus, better performance, and faster recovery. Having developed a new way to evaluate brain states, BrainCo's Focus1 headband examines 1,000 EEG features of a person in real time, and uses brain modeling to determine brain states, which then help the wearer train and develop range of movement through neurofeedback.

San Francisco-based Emotiv, founded by Australia's Monash University grad Tan Le, is engaged in developing products and research related to understanding of the human brain to monitor, visualize, measure, and understand everyday cognitive performance using electroencephalography.

Neuralink—challenging the mind–machine connection

Neuralink, the mind–machine startup launched by Elon Musk in 2016, aims to give humans more bandwidth through ultrafine threads that are woven into your brain to listen in on your neurons, or literally "record action potentials" (Musk, 2019). How much higher bandwidth? Well, the electrodes used in non-invasive methods typically max out at 1–200 because of physical limitations attaching them on top of your skull, each electrode perhaps able to track 500 neurons at once. Neuralink has been quoted saying it could eventually capture 1 million simultaneously recorded neurons (Urban, 2017).

The threads, which are layered upon a 4 mm square chip (the N1), would be implanted by a purpose-built robot together with a neurosurgeon. Each wire is thinner than a human hair. A person might have as many as 10 N1 chips implanted, each with the capacity to connect to 1,000 different brain cells. The chips would connect wirelessly through a wearable hooked over the person's ear and would operate over Bluetooth. The proximity of maximum 60 microns away from each nerve ensures the ability to detect individual impulses. The company has applied for human studies to begin in 2020. During August 2020, I watched with great interest in real time as three (seemingly healthy) pigs with Neuralink implants were presented to the public (CNET, 2020). On the screen, we could see how a computer was able to model the next movement of the pigs' legs in real time, which was kind of cool but of unknown significance in the big picture. The first humans to use the technology would be those with quadriplegia due to spinal cord injuries (BBC Science Focus, 2019).

The ambitious tasks awaiting brain–computer interfaces include telepathic communication, superhuman intelligence, additional senses, simulated experiences, digitization of human consciousness, and merging with artificial intelligence (Singh, 2019). Wow, quite an agenda.

BrainGate's research consortium is developing tech to reconnect the brain to lifeless limbs using an Aspirin-sized sensor containing 100 electrodes. Thanks to an FDA investigational device exemption, the consortium has tested the device in humans (in the BrainGate2 trial).

In the next instance, medical robots—likely nanobots (in size)—might go inside our bodies and inside our brains. Most experts consider the true disruptive impact of nanotechnology at scale will be next decade's challenge, so I'll leave it for my next book, which will go further into the future than a decade.

Should you worry about brain implants or celebrate them?

What exactly are these companies developing and what will the impact be? Should you, as a professional who cares about technology, care about these developments? The answer is: Oh, yes. Why? Because they will gradually change what it means to be a human being, which we should all care about. It could also save your mother or yourself from the fate of Alzheimer's and a bunch of other neurological diseases and potentially a plethora of severe psychological conditions, such as schizophrenia, which could be a life changer, in addition to conditions as diverse as paralysis, epilepsy, blindness, hearing disability, and more.

Some of these conditions already deploy brain stimulation but it is (usually) not through a microchip of sorts, which is Neuralink's approach. Musk seems to be motivated by a frustration with the "lossy" communication that currently characterizes the relationship between humans and machines (e.g. cell phones or computers) due to the necessity to use our lossy senses (hands and ears) instead of a direct, wide pipe link (e.g. a broadband connection or actually more directly through hardware to the brain) as well as a fear that purely machine-dominated AI would rapidly become so advanced that without merging our capabilities with these technologies we would somehow "get left behind," again, whatever that means exactly.

What does it mean to achieve symbiosis with AI?

The stated aim of Elon Musk's Neuralink is "to achieve a symbiosis with AI." Again, without access to exactly what Neuralink is up to, all the observer can do is acutely watch and await the company's announcements. Those that work within the FDA have a far more significant responsibility because they get early access to what is being attempted. Like it or not, we will have to trust a combination of Neuralink and the FDA right now.

Conversely, it could also bring about some sort of human/machine intelligence augmentation, at least to the factor of 10× to 100× (or slightly more) on some tasks such as root memory and recall as well as parallel computation of strategic alternatives based on a bunch of models co-trained by humans and machines. This would arguably change the playing field in areas such as business strategy options, factor analysis in any topic as long as the variables and underlying distributions are known, silly things such as board games, and potentially much broader fields.

However, this book is mainly about the current decade, and even Elon Musk admits that due to regulatory approvals and even to some extent the time it takes to make breakthroughs in technology, the uses of Neuralink will be purely experimental and constrained to patients of neurological diseases for the foreseeable future (this decade).

The man/machine symbiosis is also happening in far more mundane ways than the brain–machine interaction (BMI) approach. I am thinking of how most of us over the past two decades have first let the internet and particularly the search engine approach become our main way to access knowledge, even with the ancient medium of books, which we now consume quite predominantly through electronic forms (e-books). Second, I am thinking of the ubiquitous tool that is the smartphone, which seems to have become like the hammer was to our ancestors—the default tool for any task. What have these two technologies done to our way of thinking? How will they evolve? Smartphones and the internet have created an online reality where we already spend most of our time (Segran, 2019). What they have done beyond that is highly contested. The most visible result may be the time spent on such devices, which amounts to several hours a day for most users. However, arguably the information access has also brought about changes in how we conduct ourselves when confronted with professional expertise. The impact is positive in that expertise can be questioned but negative in that it's easy to think of yourself as an armchair expert and far too easy to spread conspiracy theories that cannot be easily debunked given their viral life and constant resurgence online.

Third, there is the emerging, embryonic medium of augmented reality, which might end up being delivered through its own interface which looks like neither a computer nor a smartphone, due to the type of communication it purports to support. AR makes use of advanced, immersive visualization techniques and technologies that require high-bandwidth, sensory interfaces (of which there are many experimental ones at the moment), yet still will require the kind of momentum that the aforementioned tools have had in order to start progressing at a more rapid pace. However, once the product market fit is somewhat better, AR has the promise of shifting all the visual communication that we had no idea could be fruitfully virtualized onto a digital interface. That would transform architecture, art, and suchlike, but also potentially the value of communication at a distance for any use case, be it social, business, or governance.

The reason is that we would literally be able to "be in someone else's shoes" and achieve some kind of digital empathy, capable of experiencing (more or less) what the person who is in the room does. Imagine interacting that way with a prisoner, someone trapped under an avalanche or earthquake, an astronaut in space, or any other extreme situation for which you cannot easily imagine the ramifications.

Human enhancement is proceeding fast

Taken together, the work in strands as remote as human–computer interfaces (AR, MR, VR, sensors, neural links), human–human mediated interfaces (using both digital and neural stimulation), and even interface design, including designing for a future with attentive interfaces that don't require users to do work (and don't require co-presence) in order to connect and feel things deeply and intensely, is progressing at breakneck speed.

There is reason to believe that fairly extensive progress would be made in each of these fields in this decade alone, given that the underlying technologies are quite readily established, the bandwidth is increasing, the cost of computing power is going down, and now there is a COVID-19-created use case of enhancing remote work to unprecedented levels of functionality, feasibility, and simplicity.

Coupled with that will likely be a level of government R&D and innovation investment that could top that which was put in to create the internet and certainly beyond that which has been invested in cancer research, to take two postwar sci-tech innovations of previously unmatched scope and importance.

Setting boundaries for machines

The most obvious activity right now is screen-time monitoring for your kids. Why are we so engaged in this pursuit? I think because there is a perception that the screen is a passive medium, which is in itself a problematic statement given the very active roles kids can take (and learn from) during advanced computer games or Zoom conversations with teachers and friends. It is not as simple as separating educational from leisure use. Most learning happens best in a playful context and so the claim could be made that unless the setting is playful, nothing is being learned.

Yet, it seems pretty obvious that the tethering to screens most of us have grappled with over the past years due to COVID-19 perhaps is net negative.

Stewardship and ethics of machines

While the immediate challenge is how to tweak and work with machines in the everyday and at work in order to get the most out of them, there are intermediate challenges way beyond the utilitarian benefit they may convey.

Stewarding machines, even in this decade, will entail making choices about which types of machines to foster, governing their interaction, potentially limiting the co-mingling of machines that over time could develop strong machine-to-machine affinities that could interfere with human ethics and behavior, and a host of other issues. This is not to say that I think that AI will take over the human race in the short term. Nothing is further from my belief. In fact, I think the real issues in general AI are many, many decades away, potentially hundreds of years away. There is even some reason to believe that the current AI paradigm (machine learning) will run into some fairly obvious limitations which will take the air out of the balloon for more ambitious things such as artificial general intelligence (and some of us are less worried about that taking some more time to take hold). However, there are many issues that do require attention in the medium and even the short term of the next few years.

The man/machine issues that I am concerned about are things such as the appropriate incentive structure for machine-enhanced knowledge workers, e.g. fairly accounting for the role of the human vs. the technology the human is empowered by. For example, if a trader on Wall Street, using a machine, tweaks an algorithm that engages in enormously beneficial or, the opposite, horribly damaging trades, who is to be rewarded or blamed, the man or the machine, or both? What should the consequences be? What if we cannot properly account for which is responsible for which because the two are so intertwined? Should this type of integration be allowed in mission-critical social functions such as the economy (e.g. the running of sovereign wealth funds which belong to the people of a nation state)?

What are the ethics of machines apart from the ethics of those who built, trained, own, or use them? How to distinguish between these roles and responsibilities? Let's say Alexa contributes to something horrible happening in people's homes, say, a pattern of discrimination based on the answer it gives to common and contested questions perhaps on social justice, gender identity, or climate change. Is Amazon responsible? Are individual engineers to blame? Is the person who purchased the device liable? What about the user who keeps asking the question and contributes to answers? Or the people who trained the device? This may seem contrite now, but it will not seem like that even five years from now, as Alexa or its cousins or children will increasingly take on key informational, epistemic, or even educational roles in the home.

Human studies using invasive technology are typically constrained by small data sets (few patients) and are rarely aggregated to reach appropriate statistical power (validity and reliability). Until the point such studies are able to represent the wider population, which will take regulatory flexibility, new experimental setups, financial resources, and more, we need to be extremely careful with coming to broad-stroke conclusions from initial experimental findings.

Munyon (2018) has proposed a neuroethical framework that classifies technologies into restorative, augmentative, or disruptive, with quite distinct considerations for each. Restorative applications should generally be prioritized, he says, and notes that the military veterans ensure America has a high-spending threshold for such exploration and that it is clinically important. Augmentative applications he puts into the military context and his main worry is if they fall into the hands of an adversary. Disruptive applications he is slightly worried about given that they historically have occurred in places like Iran and China, both of which are seen with suspicion from a US vantage point. His worry would seem to be the use case of prisoner interrogations that violate the Geneva Conventions, as well as medical ethics.

These distinctions are nice, but they only tackle the surface questions. The ethical perspective we need to apply and evoke has to involve the cautionary principle—the idea that even if we do not at this point know of any adverse consequences, we will have to look harder to imagine what such negatives might be and potentially even stall some developments until we as societies can rule them out or at least feel confident we could reverse course, should such a consequence be discovered. For many synthetic biology applications, that option would mean keeping things in the lab longer (not stopping research), but this would have funding-level implications. For neuro-AI it might mean constraining the human experimentation slightly until it has been tested on animals for years to come, or at least constraining the experimentation to dire neurological cases for a while longer and possibly beyond this decade.

At the same time, the augmentation of human potential needs to be seen in an emancipatory light, even if the result is a kind of augmentation that produces greater differences among us, also at the level of physical, emotional, and cognitive capability—which, in turn, will have consequences for physical strength (and destruction potential), equality, equal opportunity, and earning potential. The argument Hugh Herr is making is that the policy drivers around augmentation technology should focus on enhancing human diversity (Shaer, 2014).

At the moment, the biggest issue surrounds how to make insurers (private or public) pay for a $40,000 bionic limb, but in the near future the issue may also be whether to accept that this bionic limb (perhaps an arm) will enable the wearer to have superhuman potential with consequences the likes of which we have seen in Hollywood movies—the latest I can recall would be the villain Killmonger in the 2018 blockbuster *Black Panther*. As with all technologies, we will have to think not only of the manufacture but also of the use of technology with destructive potential. Having said that, I would be willing to live with people walking around with bionic arms paid for by government healthcare as long as they had no weapons of mass destruction attached to them, if such a distinction can be made going forward. Time will tell.

However, knowing how R&D happens, how politics works, how money talks, and also believing in and observing in real time the innate curiosity of the human mind, as a realist in matters of disruptive change I don't foresee that the cautionary principle will win through in all jurisdictions. The likelihood of making catastrophic mistakes is very real and has to be modeled into our scenarios. We need plan Bs and we need to prepare for further adversity from an environment and a human society that we cannot fully control. On the positive side, I guess, it will mean a decent futurist will not be out of a job.

How brain–computer interfaces are transforming everything

The brain, with its 100 billion neurons and 100 trillion connections, remains a mystery for humankind. While the popular obsession with brain–computer interfaces (BCIs) seems new, this type of experimentation has been happening since the 1970s. BCIs have been used to restore functionality to people affected by neurological diseases, brain injury, or limb loss through neuro-prosthetics. In the late 1980s, scientists in France inserted electrodes into the brains of people with advanced Parkinson's disease, a treatment that was approved by the FDA in 1997 (Drew, 2019). Deep-brain simulation (DBS) has since also been approved to treat obsessive compulsive disorder and epilepsy and is being explored for use in mental health, notably for depression and anorexia. The challenge is that it affects personhood in a slightly more permanent way than drugs do and has been known to cause unrelated personality changes or drastically changed behaviors or outlooks on life, which leads neuro-ethicists to question whether the very notion of human agency is threatened by such developments.

So while application of BCIs is not new, the past decade has seen an accelerated rate of progress due to a chain of developments in science and technology, such as AI, robotics, sensor technology, miniaturization and IoT, a myriad of startups pursuing innovative new products (Neuralink etc.), regulatory advances (including the first FDA-approved digital therapeutics), as well as favorable social dynamics (general support for medical interventions in seriously ill patients). Before we move further, I wanted to explain how the wider field of neuro-engineering (of which BCI is a promising new direction) got started.

The birth of neuro-engineering

Neuro-engineering is a set of biomedical engineering techniques deployed to understand, repair, replace, or enhance neural systems, and is poised to grow significantly in the next decade, but is typical polymath territory that will take collaboration between engineers, mathematicians, physicists, cognitive psychologists, and computer scientists, or better yet teams of people who each have several such skills.

The field does not advance on its own. The BRAIN Initiative (2020) from the National Institutes of Health launched by President Barack Obama in 2013 has already poured $400 million into such efforts.

Not all neural activity implies working with the brain. Peripheral nerves connected to the spinal cord in the arms and legs offer easier (and less risky) access points. Both cochlear implants and visual prosthetics have been around since the 1950s but have quite recently improved exponentially through bionics, taking the devices created by Advanced Bionics as an example. Advances in software, hardware, and biomaterials are enabling an accelerated pace of device development and progress in the field, from the stereotype of a pirate's wooden leg to today's 3D printed on-demand artificial limbs digitally enhanced with advanced sensors and with bionic functionality, e.g. resembling a biological leg's functionality.

How soon will amputees with bionic limbs outpace humans?

For a popular example, you may think of "bladerunner" Oscar Pistorius, the first amputee sprinter to run in the Olympic Games in 2012 using carbon fiber-reinforced polymer prosthetics. The bionic limb prosthesis was built by fellow amputee Hugh Herr, who heads the Biomechatronics (2020) group at the MIT Media Lab. Interestingly, Pistorius' ability to compete followed a massive debate on whether the limbs provided an advantage over human limbs, which is where the field is today—bionics is at the cusp of overtaking human capability.

Today, some artificial limbs let you function nearly as well as before. The holy grail might initially seem to be providing human-like reflexes and sensation, but the real objective goes far beyond that, extending into human enhancement through technology. Through targeted muscle reintervention, the nerves in the upper arm are rewired to twitch when the users think of moving their hand. Separate sensors attach to the skin over the top of the muscles and correspond with the intended movement (Bryant, 2019).

The visual look of artificial limbs is also making massive progress through advances in artificial skin. Altogether, it is the integration of software, the use of AI, the advances in hardware, and the ability to forge new types of skin-like materials onto such devices that provide the ability to render a near lookalike of the human form factor. Such advances can be used both for bionic limbs and for humanoid robots, although the implications are quite different.

Both the technologies and the production (beyond prototypes) and roll-out of such technologies are immensely costly and at times cost-prohibitive in today's healthcare climate, whether you are dealing with amputees within single-payer or employer-financed healthcare insurance systems.

The key role of government regulation

In 2019, the FDA's initial regulatory considerations were positive as long as providers kept biocompatibility in mind and deployed animal trials first to prove safety; the guidance is subject to change as the agency learns more and the technology progresses (FDA, 2019).

The political implications of being viewed as helping veterans is one reason this type of research might get ahead in the US as other similarly ethically complex technologies might take longer to market.

For example, throughout the past two decades at least, there have been active attempts to replace or restore motoric function to people disabled by neuromuscular disorders such as amyotrophic lateral sclerosis, cerebral palsy, stroke, or spinal cord injury. The main sensory technology deployed has been the use of electrodes placed on the scalp, a test method called electroencephalogram (EEG), or electromyography (EMG), placed on the muscles, although physically invasive methods such as electrocorticography (ECoG) or intracranial electroencephalography (iEEG)) have also been deployed. The difference is the degree of signal loss, which decreases exponentially the closer the sensor gets to the brain. EEG/EMG-enabled BCIs have been able to control drones, video games, and keyboards with thought, but so far have not gone further than that.

That has also been the regulatory justification for allowing researchers (and to some extent startups and corporations) to experiment with and conduct clinical trials with neuro-technologies.

The MIT Media Lab AlterEgo headset, a device whereby the user can operate a computer without speaking or moving their hands, purely by moving their jaw, is an example where users have been demonstrated to do math, make phone calls, order pizza, and even receive assistance while playing chess (Singh, 2019). The approach uses machine intelligence as a natural extension of the user's cognition by enabling a silent, discreet, and seamless conversation between person and machine. The wearable system reads electrical impulses from the surface of the skin, which occur when a user is internally vocalizing words or phrases.

Ethical issues with brain manipulation

The ethics surrounding brain–computer interaction have always been fraught with major issues that are not easily resolved. Socio-ethical issues include privacy, mind reading, remote control, brain enhancement, liability, self-perception, and perception through others (Waldert, 2016). Will bionic limbs connected to the internet provide adequate privacy? Is it conceivable that humans can live peacefully alongside enhanced cyborgs? How drastically will the self-perception of cyborgs change? Under which conditions will cyborgs be allowed to compete or work alongside humans?

Also, it is one thing to conduct controlled experiments on very ill patients in the confines of a hospital or a neuroscience or psychology lab, it is wholly another to allow the experimentation to go mainstream among the world's rogue scientists or mind-bending Centennial and Millennial founders, or perhaps more worryingly those actors in countries where regulatory oversight is next to none. That is, however, today's reality, given that the tech's maturity has made such experimentation possible for nearly anybody who wants to try.

Cognitive science progress over three decades

What has changed? What are the implications? To think more deeply about these issues, we have to consider the broader field of neuroscience and cognitive psychology, domains I have been fascinated by for 30 years. My father was a professor of cognitive psychology with a specialty in the study of human intelligence. I was a human lab rat in his empirical research from the time I could walk. I worked in his lab from my early teens, and I became a research assistant in my mid-teens.

I remember vividly discussing with my father, who was running the psychology department at the Norwegian University of Science and Technology at the time, how to recruit a brilliant young academic couple (Professors Moser and Moser) to the faculty. He did so in 1996, and they built an impressive activity over the next 20 years. In 2014, the Nobel Prize in Physiology or Medicine was awarded to brain researchers May-Britt and Edvard I. Moser, Kavli Institute for Systems Neuroscience, NTNU (shared with John O'Keefe). The center's many research groups are currently working to determine neural mechanisms for space, time, and memory in the brain, which will have ramifications for the future of AI, for brain–computer interaction, and for a plethora of other basic ways that human cognition works, including potentially the cure for Alzheimer's disease.

The role of cognitive maps in embodied perception

One particularly promising direction at that lab is figuring out how cognitive maps work. Cognitive maps are representations we humans tend to make when we think about things (what psychology calls cognition). The idea that such representation would extend beyond spatial navigation (we clearly learn when we walk about in an environment) stems from American psychology giant Edward C. Tolman (1948). Back in the 1940s, he had the foresight to suggest that such a thing might exist. At this time, it is becoming possible to potentially verify his hunch through scientific experiments.

In one experiment, Kuhrt et al. (2020) found that humans can indeed navigate abstract knowledge and still describe what they see in quite personal terms, which then helps them remember and better navigate the space. They tested that fact by using a virtual reality headset and a two-dimensional space of circles and rectangles, each with a unique color which they had designed for the purpose (Bellmund, 2020). The two techniques explored, physical navigation (through movement) and mousepad navigation, had similar results, although follow-up studies are needed to see whether the learning effect differs over time.

In other words, both physically and visually exploring even abstract concepts and ideas enhances learning because it allows you to construct a mental representation, a mental map of what you just learned. However, multisensory stimulation seems to be a delicate balance between providing sensory stimuli that facilitate memory and overstimulating, which causes confusion.

Embodied learning is forever

This type of embodied learning shows that brain–computer interaction, with or without a physically connected neural link, has the potential to reshape how we understand and potentially manipulate and enhance memory.

What you can use this for is to be more aware of using all your senses when you study the future of technology. Both recall and understanding seem to be far more efficient if you make use of visual and physical scaffolding (things that help you simplify, categorize, and remember). It helps if those items are related to the item you are trying to learn, but any physical layout, putting things down on paper, creating your own, personalized visualization, would seem to be very helpful in this pursuit.

It goes without saying that the implications go far beyond people with preexisting conditions. The only question is how quickly the experiments in a cognitive neuroscience lab will make it into innovation prototyping, K-12, and higher learning pedagogy, and into the workplace to enable innovation more broadly.

Moreover, it validates allowing your kids to spend time on advanced computer games, which have strong visual-spatial components that reflect corollaries in the real world. It also validates using spatially inspired memory techniques for memory and recall. The question then becomes how to steer the development in a direction that makes a positive impact on humanity and where enough people are aware of the pitfalls and can react to adjust course and course-correct should that be necessary.

Human enhancement as a broader focus

Human enhancement is a broad goal that involves altering the human body through technology in order to capture value from improvements in sensory capabilities. The human enhancement market consists of players as diverse as Vuzix, Second Sight Medical Products, Samsung Electronics, Raytheon, Magic Leap, Google, Ekso Bionics Holdings, BrainGate, B-Temia, and others, and contains areas as disparate as BCI, gene editing, nutritional approaches, AI and knowledge management software, and more. There is a major distinction between companies that operate purely external to the human body and those that interface with the body, and again between those that attempt to change the human germline (e.g. make lasting changes to our genetic offspring) vs. those that don't.

Philosophically, some of these efforts are also linked to transhumanism, a type of futurist philosophy aimed at transforming the human species by means of biotechnologies. In the transhuman perspective, which typically takes the form of secular humanism, disease or aging are viewed as unnecessary. There are approaches such as telomerase gene therapy, which is a regenerative medicine approach based on the theory that the telomeres, the caps of repeated DNA sequences at the ends of chromosomes, could be artificially strengthened so that they don't shorten prematurely (or perhaps indeed at all). That type of gene therapy is believed to lead to stemming somatic cell reproduction or even curbing cell death, which is particularly speculative yet promising. BioViva's approach, which is exactly that, is controversial and not yet approved by the FDA to be carried out on humans, at least in the US. However, the work they do, measuring the over 6,000 metabolic, physiological, anatomical, molecular, and imaging markers that together go under the collective term aging biomarkers, and making some kind of intervention based on these findings, will arguably one day go mainstream (Hunt, 2019). If not, it would take a massive regulatory and social movement effort to prohibit, given what we might learn about those markers over the next decade and beyond alone.

Implanting tech on your body for the sake of science

I happen to have met Kevin Warwick, the world's first human cyborg, meaning the first person to implant a chip into his body. He implanted the chip in his forearm in 1998 and later connected this chip to a robotic hand, forming a symbiotic relationship where he was able to control certain functions of that hand through his mind (Tsui, 2020; Warwick, 2020).

To me, Kevin impersonated an immense curiosity and bravery and was not at all a sensation seeker for the pure fun of it. As a robotics professor at a UK university (Reading), he has written extensively on his experience. At the time, there was no telling whether being connected to a machine that way would have negative physiological impact, but he seems fine.

Either way, cognitive enhancement is not a monolithic topic; in fact, a multitude of dimensions is relevant, such as the cognitive domain being influenced, the availability, the side effects, the acceptance, the mode of action as well as personal factors (Dresler et al., 2019).

The military use case

DARPA is focused on developing emerging disruptive technologies to maintain a competitive edge over adversaries which these days include genetic engineering and soldier enhancement via robotics. A recent report on Super Cyborg Soldiers 2050 enthusiastically claimed that "direct neural enhancements of the human brain for two-way data transfer would create a revolutionary advancement in future military capabilities" (Britzky, 2019; Cox, 2019). Note that this is the same type of technology Elon Musk claims will be available just a few years from now through his startup Neuralink.

Cognitive enhancement projects are nothing new to the military and have been historically pursued through both the pharmacological (modafinil, commonly used by fighter pilots, or amphetamines) and psychological routes. Neurostimulation is now more actively being experimented with for the same purpose (Seck, 2017). Combat exoskeletons have also been tried but often introduce fatigue because of their weight and maintenance requirements.

Human enhancement within sports

Elite athletes have been known to try everything (legal or not) to gain advantage. The most proven among such approaches are performance nutrition, supplements, legal stimulants such as caffeine, positive thinking, and meditation. Increasingly, performance nutrition is becoming a pastime among the global elite in a bid to stay healthy and sporty and to gain ground.

In the vaccine domain more broadly, for example, companies such as Inovio, Moderna, and CanSino Biologics are testing mRNA and DNA vaccines to counter SARS coronavirus-2.

Synthetic biology advancing—messenger RNA

Messenger RNA (ribonucleic acid) is a part of our human biology which produces instructions to make proteins that may treat or prevent disease. In fact, RNA is just as critical as DNA, but gets less press. For example, the two new synthetic mRNA vaccines under development for COVID-19 will introduce the real possibility that synthetic biology will start to change human biology. The mRNA approaches making this possible were discovered more than half a century ago but are now accelerating and moving into clinical practice. The emerging business of synthetic gene therapy will, however, go far beyond vaccines, with both breathtakingly positive and worryingly negative implications if not handled with extreme care and attention to both the science and the ethics. This is especially relevant for synthetic gene therapy, which aims to be regenerative, e.g. make lasting changes that not only block disease but potentially also alter other aspects of this important building block of our bodily ecosystem. By expressing or inhibiting genes in a target tissue (RNA interference is a discovery for which Craig Mello earned the 2006 Nobel Prize for Physiology or Medicine), a whole array of possibilities opens up, especially as nanotechnology gets deployed to carry out the work from inside our bodies. In the process, we are becoming more cyborg.

Why we develop new machines

Developers (whether software programmers or robotics engineers) are engaged in developing new machine tools to serve the contemporary and future workplaces. Computer engineers in this decade will develop robots, programs, and wearable devices that would seem to transform work, play, government, and society. Having been mostly flashy playthings for a decade or so, wearables will perhaps start to outshine cell phones already toward the end of the decade.

Robots will seamlessly integrate into our homes not just as vacuum cleaners and communication devices with the internet (Alexa, Google Home) but also increasingly as servants, disinfection agents, knowledge avatars standing in for ourselves, and perhaps companion robots to the elderly, the sick, the lonely, and the sexually hungry among us.

Software programs will similarly improve and extend their functions to encompass a much larger scope of work-relevant tasks, organizing, executing, and searching information on our behalf. Hopefully, full-fledged software suites will develop that represent technology stacks that carry out or augment the entire workflow and task chain of core business functions (marketing, sales, accounting, etc.) so that we can simplify our workday as much as give ourselves choice and range, which is today's predicament.

To an increasing degree, developers are also in touch with end users at all stages of development, which means some of us will be exposed to some of these new features at an early stage. The special thing with experimenting with machines at an early stage is that when few design choices are made, it is open ended. We can be part of imagining new use cases, dreaming up scenarios where we would want technology to help us. I can think of many.

For example, 3D printing parts may become commonplace. The process is novel in many ways, and is often called *additive manufacturing* because the raw material is all used and there is typically no cutting down, no leftover materials (no subtraction), which is good from a sustainability and speed of manufacturing point of view. A few of the companies innovating in this space that I've worked with include Formlabs, Desktop Metal and Inkbit.

However, somebody has to design the products and set up the patterns to be printed, ensure that the right ingredients are poured into the printer, that the supply chain is working correctly, and that the process goes well. If these are critical parts, humans may also be required to test products to ensure they have the desired quality and durability. This will, increasingly, also be required by law. We will surely not want to have a computer-printed reality emerge unchecked by humans.

The skills needed to develop new machines are typically reserved for professionals who have a degree in engineering. Whether this will last is another question because some of the creation is incredibly knowledge intensive and requires university-level insight, while other parts of the creation process are less taxing on the science and engineering side of the equation and this is more a coordinating and assembling function. Either way, this first role is still quite challenging and requires intense study and exposure to the elements of creation.

Correspondingly, engineering professionals need to dig deeper, study wider, and embrace the polymath attitude I with others call *Pi-shaped expertise.* They also need to be prepared to communicate with and defend their approaches and algorithms in front of lay persons, regulators, or imagined end users, who also have a right to challenge what is being developed.

At the same time, bottom-up designs will become more commonplace. Innovation will, increasingly, come from left field. To what extent are you and your significant others, your kids even, prepared to take an active part in cocreating this decade and the next?

As I write these words, I am highly conscious of the fact that my three kids, now 7, 11, and 13, will reach the workforce as this decade comes to an end and will have to grapple with a new reality where the Earth's biophysical environment is significantly degraded, where pandemics rage, and as such where a set of key technologies will need to be further developed or stewarded (a topic I will address in a second).

For example, knowing about blockchain will not just be something a first adopter needs to worry about, being able to understand and build off of its potential and its preexisting baseline will be part of society's basic infrastructure. Furthermore, synthetic biology will not be limited just to simple organisms but will be the building block for engineering and reengineering entire species, or at the very least much more complex organisms, depending on the speed of development, which it is pointless to try to forecast.

The role of technology development is perhaps the clearest path. The role of regulatory developments is more unclear, and will differ based on culturally embedded principles and risk profiles of institutional infrastructures (the notable example of the cautionary governance principle exemplified by the EU as well as to some extent by countries like Canada).

The role of business models is far murkier, as new machines will have to find their markets fast enough to thrive. The history of hardware products is full of scrap metal, and unless a perfect design thinking paradigm prevails whereby there is always a sufficient set of end users prepped and ready to adopt them, it will continue to pile up. R&D will never be perfect, and if the risk and innovation arc is high, some products will necessarily fail, otherwise the experimentation would have slowed to a halt.

Social dynamics will continue to surprise us and do not seem to be fundamentally amenable to machine manipulation, which I would argue is a good thing. Having said that, deepfakes, social media manipulation, fake news, and other phenomena will continue to threaten the public sphere. Again it will be the stewardship, through regulation, but also through social and consumer choices, that will ultimately curb tech innovations that drag our civilization down.

What should the response be? If you want to have a role in developing new machines, you need to be conscious of the short- and long-term consequences of the technology you develop. That may not be a skill afforded to engineers alone and will require deeper study by others around the development team. I think there is a limit to the specialization that will be desirable (and possible) in this new decade, and you have to choose your depth and breadth of skillset carefully and be adaptable. Recombinatory skills will produce true novelty ultimately.

Tweaking machines

Machine operators at factories, computer users in the workplace, and remote workers will have the greatest ability to tweak their machines. Tweaking machines at factories will soon be the only human jobs available, so this is truly important. How do you educate yourself to tweak a machine? Some of it has to be on-the-job training specific to each machine. Other aspects require awareness of the man/machine issues that regulate the production process, the legal requirements, the business model you are hired to fit the machine's work into, and even the social dynamics between humans and between humans and robots in the workplace. This will become especially important as some of us will take on percentagewise robotic characteristics through our cyborg functionality in terms of artificial limbs, exoskeletons, or mind-enhancing drugs, wearables, or implantables like brain implants, and synthetic biology applications like engineered blood. The entire notion of what it means to be a human will necessarily evolve over the next decade. The full integration of man and machine will not happen in this decade, but some of us, willingly or not, will get fairly deep into it.

Self-driving cars may be the most apparent example. In order for full-fledged self-driving to take place, the symbiosis between man and car will have to become more pronounced. That is in fact what is happening when we allow sensor data to reach us in real time and to affect our second-by-second actions by the steering wheel, brakes, or otherwise, with potentially life-altering impacts on us, our passengers, other drivers, or pedestrians. It is, however, not the only use case to worry about.

Other examples are repair workers, who will increasingly come fully fitted with sensors, cameras, and technological extensions and enhancements of themselves. I have worked with the Smartvid startup, which has the most advanced camera tech, with embedded cameras with AI-based categorization of images relevant to repairs and infrastructure, which in conjunction with humans can capture, organize, share, and control visual project data, achieving a single source of truth accessible both in the field and in the office. Google Glass Enterprise, the wearable, is another form factor for the same type of functionality.

Furthermore, there is, of course, the debate around the experiments with cameras on police officers, which creates an alternate reality where there is an "independent" media verification of eyewitness accounts and police behavior.

Some of us are already a bit bewildered by the amount of choice in computer equipment and software required to carry out our knowledge workday. That choice will surely not diminish throughout this decade.

Manufacturing workers will find that their industry sector again is about to become sexier as the image problem of "factory work" will disappear because the drudging, monotonous, and dangerous work increasingly will be taken over by machines, and the more interesting cognitively, emotionally, or visually challenging jobs will remain with humans. In that process, new jobs will appear because coordinating the emerging robotic reality is not trivial.

While it can be said that the privilege of developing new machines will not be granted to everyone, it is also true that as the platforms further develop we may at least be able to configure robots ourselves almost to the degree we desire. This will go far beyond selecting colors and external features of our next car. As home robots become more commonplace, their form factors will diversify and the hot-pluggable functions will become more familiar. We should soon be able to tweak the functionalities and personalities of our machines to fit our fancy, the way we currently can change Alexa's voice. What becomes important, then, is to have clear preferences and know what interacting with machines means for various members and aspects of our household. Which areas will you not let machines in on, if any? Where do you draw the boundaries for what data the machine's producer can collect?

Currently, Alexa and Google Home are building an unprecedented inventory of everyday conversations from the hundreds of millions of homes in which they are installed. Is that okay? We will soon need to decide, and should have already, as individuals, families, and governments.

The challenges extend far beyond data and privacy and into human agency, identity, authenticity, and responsibility, which slowly are evolving (and should) into veritable domains and courses of study. In the early years, from 1995 onwards "EEG biofeedback" was by far the dominant term used, but this has evolved into "neurotherapy" over the past decade or so. The hardware has moved from desk based to wearables (Clarke, 2013).

How to access advanced technology?

One relevant question might be: Where do I access the advanced technology necessary to be at the forefront of man/machine symbiosis? The reasonable question is the opposite: How can I be prepared for when I or my kids get exposed to man/machine issues and are forced to make choices? It's a bit like getting your kids ready for the reality of puberty: It comes faster than you think and once you are there, things can happen fast.

Conclusion

In this chapter, I take it as a given that the move toward deeper integration with technology will continue apace, although to a varying degree depending on exposure to the big systems running society, notably the technology sector, the entertainment industry, and the healthcare sector.

Those who have, require, or desire artificial limbs for medical purposes have been cyborgs for some time for experimental purposes and will increasingly be at the forefront of this development, whether they want to or not. Being a recipient of such innovations is truly powerful. Taking a role as a learner or teacher of what humanity's next phase (the man/machine integration age) is likely to entail is powerful. Experimentation can also mean liaising with those who have such direct experiences in order to learn from them.

Digital therapeutics has now been approved by the FDA and is impacting oncology and other diseases. Those with neurological diseases will, similarly, soon become the guinea pigs for the next evolution of bio-cognitive technologies when neuroscience moves to its experimental apex with physical mind–brain integration through implanting chips in the brain.

This development will not conclude during this decade, but it will undoubtedly begin rolling out its experimental phase, depending quite a bit on when and where regulatory constraints are lifted. Several Asian countries (notably China), as well as some Californian companies, are likely to be at the forefront of this movement. European tech will be hampered (or reigned in, depending on your viewpoint) by ethical and regulatory developments that are built on the cautionary principle.

The best option, which gives you the deepest engagement with technology and the highest cognitive rewards, is the man/machine approach: actively seek integration with various available machine intelligences and hardware. Make sure to spend enough time training the most advanced AI you can get hold of; everyone else who is smart will be doing the same.

The caveat is that this is the riskier approach. You may face backlash. You may be exposing yourself to influence that cannot be reversed, and you may be physically harmed in the process. The rewards, however, could be significant. We all need to weigh our options.

At the end of the day, do I always recommend becoming a first adopter of any new technology that promises to enhance a human function or bring us closer to a symbiosis with machines? Not at all. There are significant risks involved. You could also waste valuable time and spend needless financial resources just to have the satisfaction of being an early adopter with the cachet that at times follows such a branding choice. The quest of the futurist is to take that on, but even as futurists we have the choice over where to invest our time. All that glitters is not gold.

Being an informed second adopter, or better, an early adopter but not the first, is usually a better way unless you have a personal conviction or an explicit medical or commercial goal to reach. Either way, this is not really about adopting innovations at pace but about stewarding humanity toward a better future, or at least toward a future that sustains the human race. The future of technology as regards man/machine integration will be full of pitfalls as well as breakthroughs.

Key insights and takeaways

1 What is your view on how far humanity should go in terms of merging humans with machines?

2 To what extent do you think that regulation is desirable or even possible once scientists have made discoveries or wealthy funders have made new experiments possible?

3 Which of the five key roles available to you as this decade progresses are you (a) currently the most comfortable with, and (b) the most uncomfortable with?

4 Recall that the roles are: (1) to develop new machines, (2) to tweak and use machines, (3) actively integrating with explicitly chosen machines in order to a achieve cognitive leap, (4) setting boundaries for the entry of machines into your life or those of your significant others, and (5) taking on the stewardship and ethics of machines.

5 Next try to think of your children, or if you don't have children, your best friend's children: What do you think they will be comfortable with? How can they get ready?

Conclusion

The future of technology is in the hands of human beings and is not subject simply to runaway technologies, corporations, social movements, or even environmental havoc, at least in the current decade. In my estimation, this is the time to turn things around to make sure that human beings continue to retain control.

There is no shortage of doom-and-gloom predictions about the future, nor is there a shortage of optimistic accounts of economic progress based on technological innovation and market growth. As with many phenomena, the truth is somewhere in the middle. Never have hell and paradise been presented to us as two distinct choices. I do not mean that there is somebody offering up that choice in a conventional way, but rather suggesting that given where the world is going, it would be smart to stop and think about what direction we'd like that future to take.

While I am not the first to recognize future trends, I do want to be the first to point out that it is the joint responsibility of individuals (*you*) and collectives (*you* as a social and political actor) to chart a slightly new course. This is accomplished by developing a deeper understanding of the dynamic forces at play in this decade and demands moving from a multidisciplinary mindset comfortably juxtaposing several domains (which many have by now) to a transdisciplinary mindset (being able to regularly find and exploit the core ideas in common between disparate domains).

I also think there is great merit in becoming a postmodern polymath. With increasing complexity, the need for wise counsel is greater than ever. Right now, it's hard to know where to look.

This book does not have an answer to all the specifics that will become important over a 10-year span, but I have offered a view of what I see as the most salient technologies and disruptive forces acting upon those technologies.

Be aware of macro-trends

Throughout this book, I have mapped the future of technology in two ways. In the first part I took the macro-view, and focused on the forces of disruption and the five technologies that matter. In Chapter 1, I explained how technology, policy, business models, and social dynamics act as disruptive forces and what role each of them may play in the coming decade. I also provided conceptual clarity on what the next-order distinctions are within each disruptive force, which is something few previous frameworks have attempted or succeeded with.

In Chapter 2, I looked at some specific ways in which science and technology enable innovation. As it turns out, there is really no way to engage in the future of our society without deeply engaging with those two domains. I do not mean to suggest that all of us can be experts in all sciences or can become engineers, but I do think that traditional schooling is not going to be the answer either. Instead I believe a continuous engagement, retraining, and reflection on these topics will need to become commonplace. Those who wish to truly transform each of those fields or make innovations that truly matter based on insight from those fields will have to start early and choose their expertise wisely.

In Chapter 3, I considered how policy and regulation moderate market conditions. This is hard to do in the abstract as the world is a complicated place. However, to simplify matters, I offered the view that what happens in a few specific jurisdictions might be particularly deserving of attention: the United States, the EU, and China. It is, in fact, surprising to me that teaching has not almost completely pivoted to focus on all aspects of what happens in the world's biggest economies. The reason is, of course, that the principle of proximity holds: we are always the most concerned about things near us. Despite technologies becoming more and more advanced, we as humans are still situated quite squarely in time and place.

In Chapter 4, I looked at how business models and startups upend markets. Startups have become almost an obsession of corporations looking to gain a shortcut to innovation. I know, because when I was introducing MIT's 1500 startups to industry, it was like selling lollipops to kids. Making something new is difficult but is greatly helped by creating a new structure to experiment within—a startup-like entity—rather than attempting to do something new within a well-oiled machinery optimized to do something else. The incentive to innovate is much higher in a startup than in a big entity. But experimentation with business models is not only a topic for startups. Corporations and even governments have started to realize that they cannot continue with what they were doing in the past if the world is rapidly changing. The challenge with new business models is that they are unproven and may need to evolve faster than before and in new ways.

The emerging reality might be more complicated than it seems. Many believe that platforms are the answer to any and all questions about scale and profitability, but today's platforms may become the equivalent of yesterday's "infrastructure." And from the industrial revolution onward, infrastructure became a concern of government. Government concern about infrastructure was appropriate then, and the current government concern about platforms seems appropriate now. Now that platforms have to a large extent become privately owned, governance becomes a bigger challenge. But there might also be evolving business models that are even more advanced than platforms—we just haven't grasped them yet. It is easy to forget there is always something new around the corner.

As governments lose control with society, the private sector is not necessarily stepping in with the same amount of governance. Private-sector governance, through boards and shareholder capitalism in public companies listed on stock exchanges, has traditionally worked quite differently. With new technologies emerging that need to be closely monitored, distributed, and standardized, and where negative consequences need to be remedied, it is not immediately clear that a market economy, or the current education system (MBAs who go through business schools), have all the answers. We don't know the answer yet, but we will, increasingly, find out. Individually, it has to do with self-learning, collectively it has to do with forms of governance that continuously ask forgiveness and which reestablish legitimacy on a regular basis.

In Chapter 5, I considered how social dynamics drive adoption. It seems obvious to say that consumers drive consumption, yet that statement belies the complexity of how consumers practice purchase behavior. Technologies have become embedded into the fabric of society, not by chance but by conscious choices of consumers.

It is also important to recognize that consumers are creatures of habit and are not always rational actors making conscious purchasing decisions. We are habitual creatures and those habits, once adopted, become self-fulfilling prophecies. When I make the choice of continuing to purchase smartphones from a certain provider, say either from the Apple or Android ecosystems, each time I make that choice it constrains me a little bit from reverting my choice later. After a while, it ceases to become a question.

The same is true about a plethora of social dynamics. They may start out as a distinct, conscious decision, or perhaps the opposite, as a spur-of-the-moment choice. However, as time goes by, even small micro-choices start fitting into a pattern. Moreover, technology becomes ever better at finding patterns in human behavior and capitalizing on those patterns. It is now capable of finding patterns that we do not even see ourselves.

In Chapter 6, I looked at five technologies that matter and why they matter as much as they do. Because of their interrelationship, I chose AI, blockchain, robotics, synthetic biology, and 3D printing. While I could have chosen others (quantum technology, nanotechnology, or artificial general intelligence) whose potential will become more apparent in the coming decade, I focused on these five because they will become commonplace relatively soon and so it is important to have deep awareness of their potential for your kids, or for anybody concerned about those that belong to Generation Z or particularly to Generation Alpha—those born squarely in the 21st century, from 2010 to 2024 (Pinsker, 2020). This does not mean that coming generations will fully understand them, but rather that these technologies are likely to have the greatest impact on those generations. Wisdom is unevenly spread across society, and it is usually a privilege beheld by those who have good mentors and who are boldly experimenting but who are, in the end, humble about what they can know about the future.

Act on micro-trends

In the second part of the book I took the micro-view and focused on the question of how individuals can effectively respond to disruption. These were not meant to be theoretical chapters; the objective was to encourage you to consider how *you* respond to disruptive technologies in your daily life.

The reason I added these three chapters, which are focused on how individuals respond to disruption, was to underscore the importance of how these *macro*-forces have an impact on the individual—you. To help explain, I am going to share with you an anecdote that does not necessarily put myself in a great light but is entirely germane to the premise of this book.

For some time now I have been wondering why my computer is slow and slowing down. It manifested in a few ways, but notably because the screen froze up and because I could not type—it was as if the entire keyboard was frozen. At other times, it was "almost" frozen. At first, I tried a lot of conventional advice. I reinstalled software. I closed programs. I turned my computer on and off. Sometimes it worked and sometimes it did not. I even tried pushing down harder on my keys, as if that would have an effect, and sometimes it did. A few weeks ago, randomly, I turned my keyboard around to clean it. I then discovered something I should have known. I have lost the plastic cover which holds the battery. The result is that sometimes the battery gets dislodged and, as a result, the keyboard ceases to work. The reason my computer starts to work again must be that I move the keyboard slightly when I turn it on and off and the battery tilts back into place.

I do not feel particularly wise for not having discovered that there might be a simple fix—that of some tape—or for not just turning the keyboard around to pop the battery back in. But the reason I mention this anecdote is first to illustrate that even with deep engagement and awareness of technology, it can still be challenging to see the forest for the trees. As somewhat of a computer expert, my first instinct was not to look at the keyboard but to immediately think the problem was much more complex and so look at a myriad of other factors that could cause my computer to come to a standstill. In reality, my computer is six years old and is nearing its life's end, as programs have become larger and the processor, the RAM, and the amount of storage it contains simply are not as impressive as they once were. The hard disk is also about to give out and needs to be replaced by a faster version. All of these things could be readily observed by opening up the computer's task manager where disk, memory, and CPU constantly flare up toward 100-percent usage seemingly at my slightest touch.

In addition, I have a work style where I keep an enormous number of programs and windows open at the same time. I am not great with simultaneous capacity in other parts of life, but on the computer I somehow am. I am quite efficient at it—but obviously not when the computer cannot keep up or the keyboard is not connected to its battery.

What is the lesson here? One, experts sometimes look for solutions in the wrong places by overcomplicating matters, and partly because they are right, the problem does also manifest elsewhere. However, experts tend to forget to look in the most obvious places, and forget to use their common sense, as I did in this instance. I would hope that as we all move through this first decade of the 21st century, we are able to simultaneously act like common-sense folks and as experts, taking multiple roles and perspectives at all times. In fact, the ability to be and act with role-based empathy is part of wisdom.

If we want humanity and our Earth's ecosystem to survive and thrive for generations to come, we need to simultaneously act like common-sense folks as well as experts. The answer to every question is usually not simple or singular. I encourage you to always look for many or all the explanations that are possible, and work from there, making few assumptions. This is, obviously, different from the most efficient, expert-based approach, but it is an important counterpoint. If we leave the ethics of AI to the AI engineers, where will that leave us? If we leave politics to the politicians, where will our society be? You get the picture.

In Chapter 7, I considered the important objective you should have to become an expert in two domains and well versed in dozens. I argue you should aim to become a postmodern polymath who defies disciplines and knows a lot about a lot. That could mean many different things, but it certainly means that you are dangerously smart and aware of at least two subjects of some societal importance, and have a wide awareness of all other relevant societal and technological topics so you can have a conversation about them and form an evidence-based opinion. You also have to be willing to engage in dialogue, even with those who do not deploy sci-tech-based arguments or who may not even make sense to you.

This is not an insurmountable challenge, but it is not easy. It is also slightly more complicated than dedicating 10,000 hours to the problem, as is conventional wisdom since Malcolm Gladwell popularized the strategy. But if society is going to make sustainable progress, at least 10 percent of the population, perhaps even more, needs to have the ambition to become polymaths over the next decade. We cannot afford to go ahead with all the advanced technologies we are pursuing otherwise. We simply would not have the critical mass of voices needed to direct them wisely.

This has personal, economic, and political consequences. It matters for education. It matters for innovation. It matters for the personal choices you make. You need to prepare both for the worst and the best future you could possibly imagine. The stakes could not be higher. When the opportunity is great, opportunities lost can be cliffs of despair when looking back.

In Chapter 8, I tackled how you should personalize your insight ecosystem. You cannot rely on others—even mentors, teachers, mainstream media, governments, or external authorities—to inform you. Build an insight network consisting of information sources, connections, friends, and influencers who give you the input you need—and this is the truly important part—to rework everything you digest and to develop a personal perspective, a mental model, and a set of goals and actions based on what you find.

The challenge of a truly personalized insight ecosystem is that it cannot be outsourced fully. Your brain, and your embodied perception of technology and its consequences and opportunities, are just that—yours alone. You should team up but avoid expecting others to have the answers. To act appropriately without getting paralyzed by choices is the essence of being human. We need to learn to fail to be resilient, getting up again and still being prepared to fail yet again. That is the scientific process that has ensured progress is made. It is the only game in town.

In Chapter 9, I reflected on how you can take an even bigger leap and merge with technology to achieve a cognitive leap. The expression might seem a bit strange. After all, what does it mean to merge with technology? What does it mean to take a cognitive leap? Yet, it has happened slowly over the past few decades to more people than I can count in this chapter—those with hearing aids, artificial limbs, those who pursue careers in computer science, in robotics, and now frontline workers on the factory floor. It is even happening to knowledge workers, whether in traditional professions (lawyers, doctors, engineers) or in white-collar work. It is also, in fact, happening to blue-collar workers as they are being augmented (not just replaced) by robots, 3D printers, and various other operational control systems that are flooding the factory floors. Unfortunately, politicians have not had the same chance as other professions to understand sci-tech, which leaves them stuck in perspectives that are several years behind where the real debate should be. This is unfortunate, but it is fixable. We need to allow for that fix to happen. All of us need to help.

For those groups who have deeply engaged with technology and have let it become part of who they are, part of how they express themselves, how they learn, how they play, and how they live their life, the next decade is exciting but also scary. We have a sense of what is coming. We know things have already changed significantly. We have, ourselves, struggled to keep up. We have not always made active choices, we have let things "happen." After all, is it not in the social realm that we have let smartphones completely take over? We let apps decide which dates to go to, which games to play, we pursue our hobbies online, and now because of COVID-19 we even substitute hugs and kisses with our loved ones for virtual interaction.

Understand the stakes of missing the picture on tech

At the end of the day, the future of technology will always be full of surprises. That is the nature of the future and that is the nature of technologies. What we can do is prepare for the unexpected by at least preparing for the expected and accounting for how we would react to various X factors. We can, in fact, train for the unknown by simulating the future through scenarios that might not match the future but surely will contain aspects of it.

A good start, in my estimation, is to take action on personalizing your insight ecosystem and designing a life where you continue to learn yet are not afraid to speak up based on limited knowledge if you feel that is called for. In that process, some historical perspective based on consulting old masters of the social science canon, including Marx, Weber, and Durkheim, is not wasted. A lot of their insights stemmed from close proximity to the industrial revolution, a period in which as much was up for grabs as it is in the period within which we find ourselves today. As always, reflection is often at its most profound when confronted with great danger and possibility. However, historical reasoning always has to be complemented by considering contemporary theorists. I would offer Bourdieu, Piketty, Harari, or even the field of science and technology studies (a myriad of authors come to mind) as examples of such additional perspective.

If you retain only one thing from this book, it should be this: A set of technologies is not just a means to an end. Technologies are ends in and of themselves. When a technology is created, a whole set of instructions follows. You can, at times, choose not to follow those instructions, and you may end up innovating, or simply failing at using the technology for anything useful. What you cannot do is assume that it is simply innocently there as a potential that you somehow have to engage with. You have choices. You could even choose to alter technologies to fit better with the kind of society you want to foster. But in order to do that you need enough foresight to know what is happening. You may not need to, or be able to, stop a technology before it is out of the gate, but you should for sure be able to stop it if you see it hitting something valuable. However, the insight required to intervene is more than observational skills, you need to operate the technology in order to know what it does; you may even need to be a bit of an engineer, you cannot stay at the service level. That's the kind of awareness we need among the leaders and the citizens of the 21st century.

Humans are not just containers or vessels, we are actors. Uncertainty should never stop us from expressing opinions and refining those opinions as we figure out more. As you continue that journey toward a greater grasp of technology, share your insight and your journey with the community of readers of this book. I'll meet you at Futuretechbook.com.

APPENDIX

Types of platform technologies

Biology platforms

Science labs (19th century), the Human Genome Project (1990–2003), DNA
sequencing (1970s), CRISPR (1987), NGS (2000–), Open Source Pharma
Foundation (2014)

Content distribution platforms

Storychief (2017), Contently (2010), Percolate (2011), Sprinklr (2009),
TapInfluence (2009), Kapost (2009), NewsCred (2008), Brafton (2008),
Curata (2007), Scoop.it (2007), Hubspot (2006), Skyword (2004)

Content management systems

Wordpress (2003), Drupal (2001), Sharepoint (2001)

Data mining and harvesting platforms

KNIME (2006), HP Vertica (2005,) Sisense (2004), SAS (1976), SAP (1972),
Oracle (1977), IBM (1911)

Data visualization platforms

Bokeh (2013), Canva (2012), D3.js (2011), Domo (2010), Tableau (2003), Excel (1985),

E-commerce platforms

WooCommerce (2011), BigCommerce (2009), Magento (2008), Wix (2006), Shopify (2004), Squarespace (2003)

Future of work platforms

Cloud storage and file sharing: Google Drive (2012), iCloud (2011), Google Cloud Platform (2008), Microsoft Azure (2008), Dropbox (2007), AWS (2006), Box (2005)

Customer Relationship Management: Insightly (2009), Pipeliner (2007), Hubspot (2006), Wrike (2003), Salesforce (1999)

E-learning platforms: Teachable (2014), Kahoot! (2013), Thinkific (2012), Udemy (2009), Kahn Academy (2008), Duolingo (2011)

Expert networks: Newton X (2016), Catalant (2013), Alphasights (2008), Third Bridge (2007), Guidepoint (2003), Coleman Research (2003), GLG (1998)

Freelance jobs: Upwork (2015), TopTal (2010), Fiverr (2010), Freelancer (2009), 99Designs (2008), Guru (1998)

Newsletters: CNN's 5 Things, The Skimm (2019), NextDraft (2011), BuzzFeed, Morning Brew (2015), The Daily Beast (2008)

Productivity software: Microsoft Office (1989), Adobe Creative Suite (2003), Google Docs (2005), Zoho Writer (1996)

Project management software: Monday.com (2012), Asana (2008), Smartsheet (2005), Wrike (2003), Basecamp (1999)

Team collaboration software: Trello (2011), Slack (2009), Yammer (2008), Jira (2002)

Videoconferencing: Zoom (2011), GoToMeeting (2004), Skype (2003), Facetime (2010), Cisco Webex (1995), Microsoft Teams (2017), UberConference (2012), Cisco Telepresence Manager (2011), TANDBERG (1933)

Hardware platforms

Smart devices: Samsung Galaxy (2009), iPhone (2007)

Video game consoles: Xbox (2001), PlayStation (1994), Nintendo (1985)

Mainframes: IBM T-Rex (2003), IBM (1964), IBM 700/7000 series (1952–), UNIVAC I (1951)

Memory chips: Samsung SDRAM (1992), Intel DRAM chip (1970)

Microprocessors: AMD Athlon (1999), Intel Pentium PRO (1995), Intel 4004 (1971)

Commodity ICs

Supercomputers: Folding@Home (2020), IBM Summit (2019), Beowulf (1994), Cray-1 (1976), CDC 6600 (1964)

Manufacturing platforms

Augmented reality: Hitachi Lumada (2016), PTT ThingWorx (2009), Upskill Skylight (2010), NeoSensory, WaveOptics, Wikitude, DAQRI (defunct)

Additive manufacturing: 3D printing (1983)

Distributed manufacturing of parts: 3D Hubs (2013), Xometry (2013)

Factories (1608 in Britain, 1790 in the US)

Industrial analytics: Arundo

Smart manufacturing apps (Tulip Interfaces, 2012)

Industry 4.0

IoT: ABB Ability (2017), Siemens Mindsphere (2017), GE Predix (2013), Software AG Cumulocity (2010)

Operating systems

Googless Android OS (2008), Apple iOS (2007), Apple macOS (2001), Linux (1991), Microsoft Windows (1985)

Social media platforms

Facebook Messenger (2016), TikTok (2012), Snapchat (2011), Instagram (2010), Pinterest (2010), WhatsApp (2009), Twitter (2006), YouTube (2005), Reddit (2005), Facebook (2004), LinkedIn (2003)

Software development platforms

Java, .NET, Adobe AIR, Eclipse, Zoho Creator, Linx, Atom, Cloud 9, GitHub, NetBeans, Bootstrap, Node.js., Salesforce

Utility platforms

ABB Ability (2017)

Smart meter platforms: Oracle Opower (2007)

Smart home: Google Nest (2010)

Smart grid applications: Grid eXchange Fabric (GXF)—formerly the Open Smart Grid Platform

NASA's Readiness Levels

Speaking about prototypes, a taxonomy worth mentioning is NASA's tech-readiness levels, which is much more operational and short term than the years involved in our tech embeddedness taxonomy. The Technology Readiness Levels were conceived at NASA in 1974, defined in 1989, and extended from seven to nine levels in the 1990s. NASA's concern is very clear: The purpose is to evaluate technologies for a possible role in space flight. The emphasis is on moving a technology along a curve to develop trust and maturity, starting with a peer-reviewed principle.

Tech maturity

TRL 1–Basic principles studied (a peer-reviewed article)

TRL 2–Formulated technological concept (feasible and high benefit)

TRL 3–Experimental proof of concept (results on a few parameters)

TRL 4–Laboratory-validated technology (test performance)

TRL 5–Technology validated in a relevant environment (fits scaling requirements)

TRL 6–Technology demonstrated in a relevant environment (tested on realistic problems)

TRL 7–Prototype demonstration in operational environment (prototyped in operational environment)

TRL 8–Complete and qualified system (flight verified)

TRL 9–Real system tested in an operational environment (space flight w/ mission results)

Reference: NASA (2020; 2012): Tech Maturity Scale

First off, let me address an immediate concern that this framework is overly complex and perhaps outdated. Yes, it is complex, and it can perhaps be simplified without losing much, especially by folks launching technologies slightly less consequential than rockets in space. However, as technology grows more and more complex, verification of technology will become more and more important. Even your mainstream business software would have to go through a much more serious assessment of its maturity and functionality with much greater transparency than today. The reason is what has come to light regarding the inner workings of neural network-based machine learning approaches. Even the scientists themselves don't know how the results show up. That has been called the explainability problem in AI. Problems associated with this challenge include diversity, risk, discrimination, and fault tolerance.

This maturity scale has gained widespread adoption in any sector that runs large-scale technology trials. For example, it is used in the aerospace industry. It was recently recommended for complex US government procurement in defense, transportation, or energy infrastructure, or for nuclear energy systems (GAO, 2020). It is used for energy assessments in the Australian Renewable Energy Agency (ARENA, 2014), transportation and infrastructure projects). However, unless you are NASA, you are not going to need nine separate steps, or at least you shouldn't. Unless you are making a rocket, you are going to collapse steps 5 and 6 as well as 8 and 9.

If you are a startup in the software industry, these guidelines must seem particularly foreign. If you believe in the lean startup method, for instance, you'll go quickly to deployment and feedback, even expecting to fail and readjust. For most of NASA's projects, this has not even been considered an option, which also explains why its projects are many, many times slower and more expensive to run than the typical startup experiment.

Even so, they sometimes fail (e.g. Deep Space 2, the Mars Polar Lander (MPL), Space-Based Infrared System (SBIRS), Genesis, the Hubble Space Telescope, NASA Helios, Demonstration for Autonomous Rendezvous Technology (DART) Spacecraft, the Orbiting Carbon Observatory (OCO) Satellite), Space Shuttle Challenger disaster). If you want to compare, read one of the Elon Musk biographies (Vance, 2017) and study his method to build Tesla as well as his first rocket. The levels of risk acceptance were miles apart.

Productivity tools

In 2016, Anand Damani, presumably a successful consultant in his own right, wrote a smart little LinkedIn article called *9 Productivity Tools Used by Successful Consultants*. It is not the only piece written on this subject. It may not even be the best one—it omits several quite crucial tools that are essential in day-to-day consulting work. However, the article has the elements of virality. What it has is brevity, it has less than 10 items to remember

(5, 7, and 9 being the magic numbers), and it speaks to a crucial topic. Which tools should we surround ourselves with, given the choice and absent any tools superimposed by employers?

Most of us do not have the luxury of not using tools imposed upon us by others and our attention is already part saturated by such tools, be it Salesforce or the internal project management systems that surround most people's work lives. However, let's assume for a moment the fortunate situation of a self-employed professional, say a consultant, who has enough clients and time to entertain paying for a few productivity tools and with enough of a network to bring those tools to bear.

According to Damani, the tools you need are Slack, Todoist, IFTTT, Google Apps, Consulting Café, Mailchimp, Hootsuite, Tableau, and Basecamp— facilitating instant peer-to-peer communication, managing task lists, digital tool integration, productivity suite, consultancy collateral, email marketing, social media message coordination, data analytics visualization, and project management, respectively.

Indeed, those tools go a lot of the way, and other lists add things like invoicing or tax software to this list, among other things. Road warrior life is also easier with a large screen tablet, a mobile projector, a handheld slide clicker to move slides forward with a laser pointer, etc. Life truly has transformed for the road warriors among us in what has been aptly called the gig economy, as we have come to redeploy a phrase ("gigs") used about jazz musicians in the 1920s. There are also specific challenges with being an independent contractor not tied to one location (sometimes not even to a home). Clearly, software has evolved to partially meet that challenge.

Interestingly, not one of these tools makes claim to be the only tool you need. All providers have realized that people's needs are different and cannot all be catered to by one tool. Still, hybrids of each and overlaps between tools exist and will likely continue to exist; very often the tools have poor interop (e.g. interoperability). The benefit, of course, is that competition keeps everyone on their toes.

Somewhat lacking is a serious study of what the contemporary knowledge worker needs to accomplish their job and a careful mapping of the pros and cons of the patchwork of tools each of us has come to use in our daily life. It is simply striking that such an important part of work life is seriously understudied. Despite the econometric measurement of work productivity, the very orchestration of (productive or otherwise) work is poorly understood. High performers are studied, but mostly from an individualistic perspective. Entrepreneurs are covered widely, but little can be learned from them, perhaps, because their approaches are so idiosyncratic, so erratic, so tied to individual choices and destinies.

Elon Musk, the inventor, has this notion of starting with "first principles." In his mind, we need to reject received wisdom, even of the kind provided by modern physics, in order to make the kinds of productivity leaps and inventiveness necessary to, after some agony, produce Tesla, the semi-driverless car, or SpaceX, the world's first commercial space flight provider.

What would constitute first principles for a contemporary knowledge worker? Drawing on the scarce literature surrounding knowledge workers, from management guru Peter Drucker's 1960s' musings onward, we could perhaps identify the following key motivations, activities, and outcomes desired.

First, an occupation strongly tied to accessing information, knowledge, and insight, not necessarily in that order. Second, a desire to apply that insight to affect change, build a product, or fuel a movement of people and resources. Third, the intent to keep innovating, learn from mistakes, or complete feedback loops based on things learned, experienced, even failed experiments and criticism. Fourth, a desire for longevity and impact, being remembered, revered, and most of all, a longing to stand out, create a lasting impression, build a monument to display and exhibit our undertakings.

What Musk would advise us to do, in this case, is to shy away from most management fads of the last decade in conducting this analysis. If so, concepts like design thinking, open source, blockchain, servant leadership, all would have to be taken with a grain of salt, possibly even outright rejected on the basis that they have entered popular discourse and most likely have lost their original meaning in a sea of copycat approaches that diminish, clutter, and confuse the matters they were intended to resolve.

Similarly, we could not judge much from the existence of software-as-a-service niches such as customer relationship management (CRM), instant messaging (IM), productivity suites, email marketing tools, social media tools, business intelligence (BI), data visualization, or project management tools. Even a cursory glance at the very preliminary taxonomies arising from listing a host of the most popular of such tools (i.e. relationships, messaging, productivity, marketing, media, intelligence) supports Musk's hunch: There is no rhyme or reason to these categories and they don't contribute much to understanding your workday or even making it more efficient, much less do they provide a systematic way to look at key issues such as creativity, collaboration, and impact.

What are they then? What do they represent? Much to my astonishment, I have come to think of the last 30 years as a parenthesis when it comes to the productivity of work. We have so far accomplished very little compared with what we set out to achieve when computers arrived on the scene in the 1970s, which is not the same as to say that we will achieve very little. Rather, it is only when we successfully discover what is achievable and what may not be that we can make true progress, and may possibly even alter what is possible.

Am I a Luddite, an outright technology skeptic? After all, I'm dismissing much of the progress the software industry claims about the past 30 years. We are all supposed to have become tenfold more effective because we have all these new tools.

Those who know me well, and after reading this book that this will include you, will realize that the picture is slightly more complicated. Having a critical look at technology does not equate with dismissing it. In fact, I've spent far too many hours embedded in technology to take that point of view. It simply would not be worth it.

Data-driven venture sourcing might make a lot of sense (Deloitte, 2015).

Only the information that matters

Table A.1 Big data management tools

CATEGORY	COMPANY
Manage Big Data (all 3 Vs)	• Large tech/analytics companies (IBM Watson) • Startups (Palantir, Maana, Dmetrics)
New data types	• Semantic analytics (Dataminr, Quid, Dmetrics) • Unstructured data (Meltwater, Fullintel, Epictions) • Visual or images (Graphiq) • Natural language (Quill/Narrative Science)
Identify links and concepts	• Summarization (Agolo) • Discovery platforms (Yewno, Meta, Bottlenose) • Search engines (Google Knowledge Graph, Wolfram Alpha)
Easy-access human intelligence	• New tools @traditional consultants (PWC DeNovo) • On-demand research/consulting (10eqs, Catalant (on demand), Zurush) • Expert networks (GLG, Maven, aConnect, Experfy, Kaggle) • Innovation platforms (Presans, Innocentive)

RSS feeds

Consumer search engines are hit or miss, inconsistent, and may soon become irrelevant for industry professionals. However, specialty research tools meant for academia or enterprise are hard to access and cumbersome to use. At the same time, publishing is exploding and the effect is information overload. A single search term can return a billion results, but what you really need to know is which ones are the "best" for you.

A basic starting point for becoming more conscious about the reading choices online is to use an RSS feed reader. RSS feed readers provide readily available content for conscious online readers or those who wish to accelerate their research, marketing, and sales. Feedly's reader app caters to knowledge workers with the launch of boards, notes, and annotations. You can choose between four layouts to skim through the news. The challenge with the RSS reader approach is that it tends to lead to fairly conservative reading habits compared with the discovery-based approach that the search engine interface caters to.

Table A.2   Productivity tools

CATEGORY	COMPANY
Environment–sociocultural	Trendwatching, Trendspottr
Environment–science and technology	Scientific (Yewno, Meta, Semantic Scholar) Traditional Research and Consulting (PwC, Gartner)
Organizations–competitors	Owler, CI Radar, Bottlenose
Organizations–Startups/VCs	CBInsights, VentureRadar
Customer–profile, behavior, etc.	Bottlenose, re:infer
Customer–feedback	Synapsify, all social analytics tools
Business metrics	Domo, Aviso (Sales)
Legal	Doxai, Legal Robot, okira
All news/media	Media Monitors (Meltwater, Fullintel, Epictions, Zignal Labs, Bottlenose)
Market intelligence	Bottlenose

For an approach somewhere in the middle, discovery engines (Diffeo, Quid, Yegii, and Yewno).

Discovery engines—the next generation of search engines

Imminently quicker and more intuitive to use, with the choice between visual and textual browsing and with a transparent information credibility filter (Y score), Yegii is a crossover between a search engine (Google, Bing, Yahoo, DuckDuckGo) and a research database (Cengage, EBSCO, LexisNexis, ProQuest, RelX, S&P Global, etc.).

The market for online content discovery tools will only continue to grow as the complexity of navigating online content increases. The careful professional reader of content has to assess the reliability and validity of content, but must also decide what is the best value for money, what is available in a timely manner, and what would be trusted by the target audience. The wider potential of a deliberate discovery strategy lies in achieving far better outcomes than those who only apply readily available approaches to sort through content.

In future, a lot of reading will be outsourced to machines. As this process unfolds, those who are actively experimenting with what is possible and not will be the ones setting the tone for this emerging future.

10 thought leadership events that matter

During the matchmaking events, which I labeled Startup Showcases when off campus and Innovation Workshops when on campus, we typically covered innovation models, technologies, collaboration patterns, partnerships, and trends. It would have been tempting to stick with just science and technology presentations. After all, that's what the Office of Corporate Relations and its 75-year-old Industry Liaison Program (ILP) have done for the good part of these years. They were always keeping the same format: the 45-minute academic talk, always mostly a rehash of lecture notes, perhaps with a nod or two to industry. It worked for a long time because most distinguished MIT faculty, or at least the ones invited to stage, are intrinsically good presenters, and are hired because they already have a good sense of what in their research matters to industry applications.

However, it did mean that for many such speeches, the message went far above the heads of the audience. I recall Neil Gershenfeld's academic innovator talk entitled "The Third Digital Revolution: From Bits to Atoms," which he gave at the 2016 MIT Startup Showcase San Diego (http://startupexchange. mit.edu/startupexchange/html/index.html#viewOpportunity/121). Professor Gershenfeld, Director, The Center for Bits and Atoms, MIT and Founder of Fab Lab, is a brilliant scientist. He is also a great speaker. However, his arguments are complex. He proceeded to explain how "computer science was the worst thing to happen either to computers or to science," an intricate issue in which he essentially argues that we took a wrong path and that's why computers are as they currently are, mentioning how "bits can transport mass" and other esoteric experiments, and observations like "the head is distinct from the tape" in terms of memory vs. processing engine in Van Neumann's paper on the EDVAC computer. He says this is a hack that was used to tweak Turing's machine, which was a thought experiment to actually work as a machine and was more of a hack, a hack we are still living with in today's computers, even laptops. To Gershenfeld, the objective, instead, should be to make software look like hardware.

The narrowness of expertise

Who, working in industry today, knows the latest synthetic biology scientific methods, cybersecurity protocols, artificial intelligence algorithms written in mathematical notation, or even Einstein's theory of general or special relativity? My bet would be: only those few industry professionals actively working in an industrial lab with a dedicated team of minimum 3–7 PhDs who are currently in an industry-academic research project collaboration, have a PhD, and who just walked out of a meeting with the technical project team. That's how specialized today's advances need to be to gain traction and make a difference.

What is commonplace to a MIT professor talking about his field is often rocket science even to a PhD in a corporate lab, and vice versa, of course. Science and technology has become very specialized, hard to track, and complex. It evolves every week, every month, every year. Some would say every day.

What I did to change that up was to apply more of the fast-paced TED Talk concept to our events. Professors were asked to stick to 10 minutes plus 5-minute Q&As. Industry was lucky to have 10-minute speaking time total. Startups got 2–8-minute lightning talks plus exhibit time during breaks. Our events were cut down to three-hour morning sessions from 9–11:30 and followed a templated sequence of programmatic elements: 2-minute context, 10-minute academic innovator speech, a dozen startup lightning talks focused on technology thought leadership rather than on pitching their business plan, an investor trend keynote followed by a fast-paced panel debate, with all of the above running speakers through a set of improvised questions around innovation, personal experience, and views about the future. I was always the MC, keeping consistency, pacing, and adding a bit of humor. It worked.

The lessons I take from experimenting with formats and then sticking to one over three years is that predictability is important. People need to know what to expect. They also appreciate a media-style, movie- or TV-like pacing—even the academics did—which is more in line with how they must experience life these days.

Events at universities and at many other places often seem stuck in some 1950s scenario, much like many non-evangelical churches are. Why this fascination with the 1950s' long, drawn-out, one-way communication style where authorities mused too passively over hour-long stretches? Walter Cronkite most famously on TV, President Kennedy from the government pulpit, and famous academics such as Noam Chomsky from their pedestals. Finally, what tech innovators want to do when they network is talk to each other.

Why go to a tech and innovation event? The major reason cited by participants is to learn industry best practices or R&D trends from academia. Picking up emerging trends months before they ever show up in a blog post, a book, or on mainstream media is another reason. Many executives, founders, and HR folks go to find talent. Events are good for customer discovery. Investors go to connect with founders. Founders go to find investors. Business development folks go to identify new partners or maintain important relationships. Strategy teams go to gather competitive intelligence. Consultants go to be seen and pick up new clients. Everybody goes to network.

Table A.3 Top tech, innovation and startup events

Top tech, inno and startup events
Large
Web Summit | Slush | SXSW | Collision | MarTech | CeBIT
Exclusive
Davos | Founders Forum | WSJ D.Live | Summit | Forbes Under 30
Creative
ad:tech | Adobe 99U & Adobe Summit | Burning Man | TED | Trend Hunter's Future Festival | Tribeca Disruptive Innovation Awards
Techy
TNW | RE•WORK | EmTech Digital | Techcrunch Disrupt | EU-Startups Conference
Corporate
Fortune Brainstorm Reinvent | FT Innovation Dialogues | Gartner CIO Summit The Economist Inno Summit | Fortune Growth Summit

Event quality—making proximity worth it

Having attended a myriad of global events catering to tech innovators over the last few years, here are some that stand out for their emphasis on community building, charisma, and content beyond networking and beyond panel stacking, which is seen so often.

I once attended a Silicon Valley one-day event with 10 panels, some of which were a mere 15 minutes long yet with 3–5 speakers. This kind of panel format can perhaps be done on TV talk shows with a professional presenter but are hardly feasible to pull off on a regular stage, due to all the interruptions, walk on/off stage, and obligatory introductory statements and all the rest of it. It was a horrible event. The panels were shallow. Everything felt rushed. It was evident that the organizers simply had stacked the panels with as many audience members as possible in order to secure attendance.

When it comes to startup innovation overviews, there is no match for the Finnish startup bonanza that fuels the massive event, Slush. From a student-driven start in 2008, Slush was recently attended by over 2,600 companies and 1,500 investors, networking and negotiating funding. With critical mass, a lot can get done in a short amount of time. Although Web Summit's

2017 event clocked 53,056 attendees from 15,000 companies from 166 countries, it may have become too big and hit the upper limits of an efficient crowd. The OurCrowd events in Israel get even bigger. I attended in 2019 and it was a spectacle more than an event. Who knows what will happen to these mega-innovation events post-COVID-19. I would have stopped attending without the virus; it's just not enjoyable. Others may feel energized by being around 100,000 people, I don't know.

This begs the question: Is there a Dunbar number for events? Dunbar's number, arising from anthropological research by Professor Robin Dunbar of the University of Oxford during the 1990s, is an estimate of the number of relationships that our minds are capable of handling simultaneously. According to Professor Dunbar, this number has remained at around 150 throughout human history, from the tiny village communities in which our ancestors spent their whole lives up to the modern age of international travel and social media (Mac Carron et al., 2016). He attributes this stability to some sort of cognitive limit to our attention span. Dunbar found tribe size ranges of 30–50, 100–200, and 500–2,500 members, which is interesting to map to the top technology events around the world.

The White House Fellows

The prestigious White House Fellows program is each year awarded to 11–19 outstanding candidates to the President for a one-year appointment as Fellows. This is at the lower end of a tribe but is a range seen in private school class sizes across the world. That amount of people builds a strong class cohesion yet has enough diversity to be an interesting group, with the caveat that each individual is outstanding.

Salzburg Global Seminar

Salzburg Global Seminar hosts programs on global topics aiming to challenge present and future leaders to solve issues of global concern through programs held at Schloss Leopoldskron, in Salzburg, Austria. Each two-day seminar will have an average of about 50 participants, given the size of the venue with 55 rooms, 12 suites, and three townhouses. The seminar I attended (on e-government) had an intimate feel and the attendees became well acquainted over that period. Indeed, the organizers aim for intimate, dynamic, and collaborative experiences that create lifelong bonds.

I prefer to arrange events with 75–100 participants, which, if the audience is the right mix of people, provides just enough novelty but does not break the (relative) intimacy. I have occasionally attended weekend seminars with fewer than 30 people that have been epic, but they are rare.

Location matters. The fabulous regional Nantucket Conference on technology and innovation, bringing together founders and venture capitalists in the US northeast and Boston area, is great because we are all "stuck" on a paradise-like island (Nantucket) for a weekend.

Founders Forum

Founders Forum (2005–), hereafter FF, started by Lastminute.com founder Brent Hoberman, is a series of intimate, invite-only annual global events for the leading entrepreneurs of today and the rising stars of tomorrow. FF describes itself as a network as distinct from a mere conference and prides itself on fostering discussion and being somewhat of a (launching/marketing) platform. The forum is free of charge to participants thanks to corporate partner sponsorship from Pictet, McKinsey, Google, Dell, Facebook, TalkTalk, HSBC, Nasdaq, Osborne Clarke, Wilson Sonsini Goodrich & Rosati, Audemars Piguet, NetJets, Diageo, McLaren, Smaato, FTI Consulting, and Dassault Systems.

FF is invite only since they "want to highlight that every guest could be a speaker in their own right." At the Meet the CEO session, some 15+ CEOs (country leaders of billion-dollar-revenue companies) stood next to their own banner stand and took short speed-dating conversations with attendees.

At FF London, there was a majority of startup CEOs, mainly from the digital sphere, with applications in UK core strength industries such as software, finance, media, fashion, and (high-end) retail. There was less focus on industrials, life science, and manufacturing.

Clearly, there is more to events than their sheer size, but it does provide the boundary for what type of interaction to optimize for, both as an organizer and as a maximizing participant.

Establishing your convening power

The salient variables that impact the quality of a tech innovator's quest for the ideal event to attend or speak at are size, participant match, location, regularity, communication tools, and content quality. The interesting thing is to figure out the priority of these factors. Each person is different, but here are some rules of thumb.

Prioritize exclusivity/intimacy or scale, unless you (or until you) can muster a 300-plus awesome audience.

Annual events have a chance of becoming legendary. Monthly events build community. Weekly events build camaraderie. Daily events become ecosystems. Hourly events have amazing predictability and reliability, like clockwork.

Homogeneous participants build intimacy but risk group think and boredom. Particularly, this could become a problem with compact, annual events. You want to build community, but you do not always want to see exactly the same people. Heterogeneous participants emphasize creativity but risk fragmentation. Any mix is riskier than sticking with homogeneity or heterogeneity.

Integrate speakers with the audience. Try to reduce the distance between the two. Nobody wants to be spoken to anymore. Avoid VIP or speaker lounges.

Never skimp on content. Check all the presentations beforehand. I have gotten into trouble for giving critique to C-level speakers on their content, particularly if their marketing department has had a hand in the slides, but it is essential to do this. If you let marketing slides pass, you have broken the trust between speaker, organizer, and audience based on an implicit promise of authenticity.

The audio must be impeccable. Recent research suggests that audio quality is very significant in terms of people's reception of a message.

Pacing is everything. Up-tempo but unstressed is optimal.

Resources for artificial intelligence

Artificial intelligence

The term "artificial intelligence" typically refers to a level of machine learning that equals or supersedes human intelligence. Not all AI necessarily relies on machine learning. Notably, all AI so far has been narrow AI, meaning it supersedes humans in very specific tasks, such as image recognition and a few other types of domain-specific pattern recognition.

In contrast, there is the lofty promise of general AI (AGI), which would be a type of intelligence that would match or supersede humans in a more general sense. What that might mean is entirely unknown. Whether and when computers ever will get to that point is the subject of some speculation among both computer scientists and philosophers. Some are in awe, others are deeply worried (Bostrom, 2014). As Dafoe and Russell (2016) state so eloquently and clearly: "The risk arises from the unpredictability and potential irreversibility of deploying an optimization process more intelligent than the humans who specified its objectives." For our purposes, that is, looking at technologies in the next decade, it really does not matter. We are far, far away from such a thing or from such irreversibility. But it will matter a lot in a 25-plus-year time frame, which would have to be the subject of a whole other book, since changing the time frame indeed changes nearly everything. In this book, I am concerned purely about the next decade.

Machine learning

Machine learning is a type of computer algorithms that, put together, improve from experience. The only way they do so, at the moment, is through being fed both a huge data set and an interpretation of a small subset of that data set, known as *training data*. There are a vast number of such algorithms, many of which are a legacy of statistical thinking for the last 300 years. The most notable are the naïve Bayes classifier (1763), linear regression (1805), Markov Decision Process or MDP (1906), logistic regression (1944), nearest neighbor (1967), convolutional neural network (Fukushima, 1980), recurrent neural network or RNN (Hopfield, 1982), decision tree (1984), support vector machine (SVM, 1990s), gradient boosting (1997), random forest (2001), and reinforcement learning (RL) algorithms (1960s onwards, yet also incorporating MDP such as Q-learning (1989), SARSA, DQN, DDPG, NAF, A3C, and generative adversarial network (GAN), developed by Goodfellow (2014).

Some of these algorithms derive directly from the fields of statistics or mathematics and some do not. What's fascinating is that even though the theory of how some of these algorithms work is part of our science legacy, the increased computing power of the past decade has given them renewed relevance and has led to infinitely better results. One example is how AlexNet (Krizhevsky et al., 2012, 2017) deployed CNNs on top of recently developed graphics processing units (GPUs). That fact should be reason enough to remember to consider the past as you are designing the future, or, as scientists have said before, "We are standing on the shoulders of giants," something John of Salisbury pointed out already in *Metalogicon*, his treatise on logic, back in 1159.

If you are going to execute machine learning, you need a deep knowledge of each of these in order to pick the right tool for the challenge. If you are only trying to interpret the results of an analysis, you still need a superficial awareness of how the algorithm works, e.g. what it selects to focus on and prioritizes in order to create patterns in the data.

There is now a significant and important focus on going beyond the "black box" of hard-to-interpret algorithms such as those used in deep learning, partly because it could cause known or unknown bias. For example, Amazon has pulled its face recognition algorithm temporarily from use with the police because of allegations of racial bias. IBM, similarly, has stopped working on the topic until clarification can be reached.

Deep learning

Deep learning is a type of machine learning which became technically feasible over the 2010s, relying on artificial neural networks with hierarchical representation of data (Goodfellow et al., 2016). The computer's learning can happen through supervised, semi-supervised, or unsupervised approaches. Either way, it allows computers to look for what is often called *feature hierarchies*, a type of logical structure to the underlying data. The novelty in the approach is that it does not rely on teaching the computers every detail about the domain it needs to analyze. Rather, once the basic structure is given, the approach relies on the computer's ability to structure information as it sees fit. All the user has to do is to define which features might be interesting to look at, although unsupervised learning to some extent leaves even that task to the computer.

The challenge with deep learning is that with the way it layers the analysis, it becomes quite opaque. Even experts would have a hard time explaining how the machine got to the answers it did. Despite that, it is a popular form of analysis because of the impressive results it might produce in specific domains of Big Data that become overwhelming for humans to track. Yet, machines often work with quite strange simplifications, such as defining a person by contours and corners, which is where both the challenges and the opportunities begin.

Deep learning is highly influenced by neuroscience, most notably the neuroscience of the past decade. However, being metaphorically inspired by another field to such a degree is potentially a hindrance, given that brain science might evolve without computer science fully noticing. Deep learning is nothing new, in fact many of these algorithms were developed several decades ago. The new thing is the speed of computers, which has improved (manyfold), which enables progress to take place.

Figure A.1   AI taxonomy

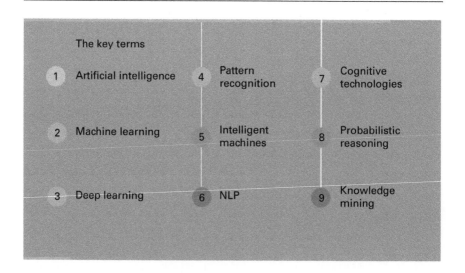

AI influencers to track

The technical field of AI is quite vast, but a good starting point is the top engineering schools around the world; preferably pick one on each continent, to avoid geographical bias. If only picking one school per innovation hotspot, I would pick MIT on the US east coast, Stanford on the US west coast, Nanyang Technological University in Singapore for its pan-Asian focus, Oxford University in the UK for northern Europe, the University of Cape Town in South Africa for the African subcontinent, and the University of São Paulo in Brazil for South America. Each is among the top ranked in its region and has an outsized regional focus in its faculty and outlook. If I could afford to also pick a Chinese university, I would track Tsinghua University in Beijing.

Another story is how to track, because browsing their websites is definitely not enough. Truly tracking scientists requires reading their papers (the most accessible ones), listening to their speeches, interviewing them, and consulting them about trends relevant to you. This is best done through an industrial outreach program such as the MIT Industrial Liaison Program, if it exists in each place, worst case by individual outreach—you might find these researchers are quite busy. Another legitimate approach is to spend more energy on more approachable universities physically closer to where you are at, which will also yield strong benefits. That may also happen if you develop a true partnership or could yield investment opportunities.

If we go into individual researcher profiles, here are some that are bound to be important in the decade to come: Andrew Ng (@AndrewYNg), who is a former chief scientist at Baidu and an adjunct professor at Stanford, who also runs the AI Fund; or Fei-Fei Li (@drfeifei) from Stanford, who formerly was the chief scientist for AI/ML for Google Cloud, and you will want to have a female perspective for sure. For Asia, tracking Kai-Fu Lee, the Taiwanese-born American computer scientist, businessman, and writer, notably the author of the book *AI Superpowers: China, Silicon Valley, and the New World Order* (2018), would be a top choice. Harry Shum, the former Microsoft executive vice-president of AI and Research, who was a key force in the development of the search engine Bing, is now an adjunct professor at Tsinghua University. For a worldwide perspective you could do worse than tracking Brazil-born, self-declared transhumanist Ben Goertzel (@bengoertzel). Transhumanism is a philosophical movement that advocates developing sophisticated technologies to greatly enhance human intellect and physiology. Goertzel has published nearly 20 books. In Africa, Moustapha Cisse (@Moustapha_6C) heads up Google's AI research lab in Accra, Ghana, the first of its kind on the continent. If you want a strong perspective on emotion AI, a major thrust in making AI benefit human communication, tracking Egyptian-born computer scientist Rana el Kaliouby (@kaliouby), co-founder of MIT spinout Affectiva, would be a great choice. Think tanks are also getting more serious about AI. Notable policy perspectives are starting to emerge from the Atlantic Council's GeoTech Center (@ACGeoTech) led by David Bray as well as Brookings Artificial Intelligence and Emerging Technology (AIET) Initiative, RAND (https://www.rand.org/), and CSIS Technology Program (https://www.csis.org/) in the US and the Alan Turing Institute (https://www.turing.ac.uk/) in the UK.

Overall, making sure you get a good mix of academic, startup, corporate, government, and non-profit perspectives is crucial to understanding the evolution of AI or any other technological change.

From the government side, note that as AI evolves with regards to both funding priorities and context-specific challenges, it will increasingly take on regional flair, e.g. Chinese AI will have particular concerns that US AI or European AI might not have to the same extent. For example, it will mirror industrial strength in particular continents and countries. Military applications seem to be at the forefront in China given its emphasis on asymmetric technology warfare, while energy concerns are key in Europe due to its sustainability focus, and overall business impact is a strong emphasis in the US due to its consistent focus on commercial application, although this is a crude simplification to prove the point which will be granular to each subspecialty of AI and will vary within a geography as well.

Top readings on AI

Tracking AI through reading books is tricky because the field moves so fast. Having said that, the Amazon list of best sellers in AI and machine learning does have the occasional find, although many books on any current list will have been published several years ago, which is usually no good in a field where one year might be worth a decade's progress simply because of two groundbreaking papers, a new labeled data set, and a new graphics card from NVIDIA. Additionally, some of these are expensive textbooks, which makes little sense unless you need them for a degree course.

Online publications such as *Wired, Techcrunch, Mashable, Ars Technica*, and *MIT Tech Review* are fertile fields of inspiration. Top individual writers worthy of tracking might include a small set of tech journalists or science journalists. Kara (@karaswisher), founder of Recode, is a Gen X influencer across tech. Since 2007, the BBC's long-standing technology correspondent Rory Cellan-Jones (@BBCRoryCJ) has sway and represents the Boomer generation's take on tech. The Verge tech writer Casey Newton (@CaseyNewton) is a much-followed Millennial.

Beyond that, my recommendation would be to closely track founders of unicorn startups in AI, which you can to through CB Insights' list, the AI Unicorn Club, which would mean you should look at the founders of Bytedance (the TikTok video platform), Sensetime, and Face++, which are selling AI SaaS, UniPath, Automation Anywhere, YITU Technology, and a few others. As can be readily understood, you will rapidly run out of time, so apply sense to this effort unless you are explicitly investing in this space. The idea would be to track the founders, not the companies, thinking that serial founders always are on to the next thing, as opposed to companies, which are forced to optimize for the business model niche they have just found.

Startups to watch in AI include Gamalon, Secure AI labs, Onfido, Interpretable AI, Affectiva, and Hyperscience.

Learning AI from tech conferences

Which tech conferences should you attend (virtually or in person) to have any inkling about where the field truly is evolving day by day, year by year. Before COVID-19, there were more than 1,000 high-quality annual industry events gathering decision makers in technology and society around the world, from SXSW via SLUSH to WSJ Tech. That number will clearly diminish over the next few years, but even the great ones will not be down to 10, not even close. While one clearly cannot attend all of them, knowing which topics are trending, who speaks at each event, which events are becoming more or less important to attend, and getting insight on geographical trends in disruption topics is important.

Virtual attendance can also potentially be optimized with clever scheduling technology and planning. I would say that the absolute top events (which may now become strong virtual events until they can fully gather face to face, but with potentially lasting improvement of their online effort) include World Economic Forum, SXSW, TechCrunch Disrupt, SLUSH, WSJ D.Live, TED, CES, CEBIT, and EMTECH. Some of these go beyond AI and digital tech and into industrial tech, but those B2B technologies definitely are covered by a different set of events as well. Browsing tech conference websites for a few minutes every month might be worthwhile. Going to the top ones you can access (within your budget) as well as a few local ones that do not require travel is usually a worthwhile annual investment, but seldom more.

Resources for blockchain

Tracking blockchain influencers

Who should you track (scientists, innovators, startup founders)? Strikingly, few academic institutions or scientists come to mind, which is somewhat worrying. Does this mean they do not track blockchain? No, it just means they have less celebrity status. Through independent analysis and a myriad of sources, I've come up with this (very) short list (XTRD, 2020).

Russian-Canadian programmer Vitalik Buterin (@VitalikButerin) created Ethereum (2013), the world's second most valuable cryptocurrency after Bitcoin, and co-founded *Bitcoin Magazine* (2011), the industry magazine. Canadian author Don Tapscott (@dtapscott), author of *Wikinomics* (2006) and *Supply Chain Revolution* (2020), is the co-founder of the Blockchain Research Institute (2017) trying to advance solutions to governance issues in the emerging blockchain industry. Barry Silbert (@barrysilbert) is the founder of Digital Currency Group, Genesis Trading (a leading bitcoin brokerage firm), and Grayscale Investments (the industry's largest digital currency asset management firm), and is one of the biggest investors in the crypto space. Neha Narula (@neha) is a computer scientist who heads up MIT's Digital Currency Initiative. Ryan Selkis (@twobitidiot) is an entrepreneur, blogger, and former venture capitalist and a founding member of the domain's largest media brand, CoinDesk. Joseph Lubin (@ethereumJoseph) is the co-founder of Ethereum and the founder of Consensys (@Consensys), the blockchain software solution provider. Kathryn Haun (@katie_haun) is a general partner at VC Andreessen Horowitz's crypto division, who spent a decade as a federal prosecutor focusing on fraud, cyber, and corporate crime. Kathleen Breitman (@breitwoman) is a controversial figure for her role as a co-founder in Tezos (@tezos), the crypto-ledger platform and currency, and is launching the first blockchain-based collectible trading card game called Emergents (https://www.emergents.gg/), in which each card has memory of where it was used throughout the game and can even be signed to prove who the collector met (Dale, 2020; Lewis-Kraus, 2018).

David Marcus (@davidmarcus) leads Facebook-initiated and already controversial Libra Foundation (@Libra) with members Andreessen Horowitz, Coinbase, Iliad, Lyft, Shopify, Spotify, and Uber. Visa, Mastercard, eBay, Stripe, and Mercado Pago all pulled out of Libra because of the controversy, leaving the initiative without a payment processer (Brandom, 2019). Facebook is spearheading a "simple global payment system" through Novi (@novi), a digital wallet "changing the way money moves," which was swiftly rebranded after the initial name Calibra, a Facebook subsidiary, became entangled in controversy ("Libra" and "Calibra"). Adam Draper (@AdamDraper) is the managing director of the Boost VC accelerator, which invests in 10 sci-fi founders a year in the domains of crypto, virtual reality, augmented reality, AI, ocean, biotech, and time travel (sic), and already has a portfolio in excess of 150 companies (Cryptoweekly, 2020).

Their Twitter handles are meant as shorthand for "try to get hold of anything these guys do or touch."

The best crypto-education can be found at MIT, Stanford, and Cornell, and internationally at EPF in Switzerland and the University of Copenhagen in Denmark, partly because the blockchain's interdisciplinary nature demands a myriad of perspectives that only a top research university with many applied labs and centers can provide, as well as a focus on both the faculty and the student level (Coinbase, 2019). Student bitcoin and blockchain clubs drive the agenda in a significant way.

If I was to track only three blockchain startups it might be Coinbase (https://www.coinbase.com/), the platform to buy and sell cryptocurrency, Circle (https://www.circle.com/en/), the payment platform, and Enigma (https://www.enigma.co/), the blockchain-based security platform allowing encrypted computation on web content and data.

Best blockchain readings

Tracking news in the space is best done through *Bitcoin Magazine* (2011), *Blockchain Magazine*, CoinDesk, Cointelegraph, and The Capital (thecapital.io), a financially incentivized social content publishing platform (CryptochainUni, 2020; Detailed, 2020).

Some notable podcasts with a weighty number of episodes on record include Pro Blockchain by Max Burkov, Unchained with Laura Shin (@laurashin), Blockchain Insider, and The Bitcoin Podcast.

Keep in mind that these things change fast, so asking experts for up-to-date top three lists is always something to do (never longer lists since you just do not have time).

Blockchain events

What tech conferences should you attend? The main online blockchain publications typically run events, which is the best way to stay up to date and meet the community. Geneva Blockchain Congress (https://genevablockchaincongress.com/), hosted with the context of that city's organizational profile with UN agencies, looks at governance, legality, and ethics. London Blockchain Week emphasizes the impact on financial solutions that can be implemented today, including the Blockchain Summit London (https://www.blockchainsummitlondon.com/). Hong Kong Blockchain Week (https://www.hkblockchainweek.net/) captures developments across Asia. Blockchain Africa Conference (https://blockchainafrica.co/) presents revolutionary opportunities on that continent. Future Blockchain Summit, the world's largest blockchain festival, is sponsored by Smart Dubai and might unlock opportunities in the Middle East (Craciun, 2020). In addition, a plethora of fintech conferences will cover the topic, such as Money 20/20 (USA, Europe and beyond) (Bizzabo, 2019).

Resources for robotics

The robot experts: who you should track (scientists, innovators, startup founders)

The biggest manufacturing companies in the world will make massive progress by creatively and extensively deploying robots in their manufacturing processes over the next decade. Watching what they do makes sense. Startups will continue to innovate but are unlikely to produce mass-market systems in the industrial environment without partnerships with the huge firms in industrial automation, notably Schneider Electric, Siemens, Rockwell Automation, Thermo Fisher Scientific, Mitsubishi, Honeywell, Danaher, Emerson, GM, and ABB. The technical leaders in those firms are usually the companies' spokespeople and may have blogs, podcasts, and interviews where you can glean what their experience is at any given moment. They also have industry associations where you might be able to connect.

Startup founders from the world's top technical universities are another source of insight.

Robotics influencers to follow include Colin Angle (@colinangle), CEO of iRobots; Daniela Rus (http://danielarus.csail.mit.edu/), the first female director of MIT's Computer Science and Artificial Intelligence Lab; Kristin Ytterstad Pettersen, professor at the Department of Engineering Cybernetics at the Norwegian University of Science and Technology and a co-founder of marine robotics startup Eelume (https://eelume.com/), a disruptive technology for subsea inspection, maintenance, and repair using self-propelled robotic arms; Per Vegard Nerseth (@PVNerseth), CEO of CMR Surgical and former head of robotics for industrial giant ABB; and Josh Hoffman (@ZymerJosh), CEO and co-founder of biorobotics startup comet Zymergen.

Robotics startups to watch in order to track the emerging capabilities of robots include CMR Surgical, a British robotic surgery company founded in 2014; Zipline, an American drone delivery company focused on delivering vaccines, blood, and medicines; and Righthand Robotics, a US startup which is a leader in autonomous robotic picker technology facilitating flexible and reliable e-commerce order-fulfillment, each with an interesting, distinct robotic use case. In 2020, the industry passed 100 million robotic "picks" (ET Bureau, 2020). Humatics, an MIT spinout, makes sensors that enable fast-moving and powerful robots to work alongside humans without accidents based on ultrawide band (UWB) radio frequencies, so-called micro-location technology, which is more accurate and less error-prone in factory environments and tunnels than Wi-Fi, GPS, and cameras.

For consumer robotics, the fate of iRobot, the MIT spinout founded back in 1990, will be instructive. The maker of autonomous vacuum cleaners (Roomba) and autonomous mops (Braava) has sold more than 30 million consumer robots in 30 years. Whether the consumer robotics market will sell 300 million or 3 billion units in the next 30 years is not clear at the moment, and consumer use cases for existing robotic capabilities are few and far between.

Tracking robot knowledge: what you should read (publications)

Government reports, industry reports on trends in robotics, trade publications are all useful. Specifically, *IEEE Robotics and Automation Magazine*, *Autonomous Vehicle Engineering*, and *Automotive Engineering* magazine are three specialist magazines to track. In terms of science magazines, *ScienceRobotics* (https://robotics.sciencemag.org/), which requires a subscription, is the one.

Exciting podcasts in this space include Talking Robots, The Robot Report podcast, Soft Robotics Podcast, and Robohub. They are interesting because they can give unique insight into hard-to-access observations about true obstacles and opportunities in the space that founders and leaders are more likely to reveal through a long-form interview than in short press clips or on their website.

In the robotics space, YouTube is another winner, for example Boston Dynamics' channel https://www.youtube.com/user/BostonDynamics/videos. Having initially been acquired by Google, the MIT spinoff is now a wholly owned subsidiary of the Japanese conglomerate SoftBank Group.

What tech conferences you should attend

RoboBusiness, RoboSoft, RSS (https://roboticsconference.org/), World Robot Summit (https://worldrobotsummit.org/en/), and BioRob are the five top events in the space. Such events make more sense to attend than traditional industry conferences for the simple reason that the physical hardware demos are so exciting and inspiring.

Resources for synthetic biology

Synbio people you should track (scientists, innovators, startup founders)

If you are tracking synthetic biology as an observer of innovation in the space and beyond, tracking 3–5 key profiles should be enough. If that is the case, my choice would be George Church (@geochurch), professor of genetics at Harvard Medical School; Craig Venter (@JCVenter), who led the first sequencing and analysis of the human genome back in 2001 and has been a luminary ever since; Jim Collins (@MITdeptofBE) (https://collinslab.mit.edu/); and perhaps Tom Knight, co-founder of Ginkgo Bioworks (@ginkgo). Tom Knight is a senior research scientist, Computer Science and Artificial Intelligence Laboratory, MIT, and is arguably the "father" of the field of synthetic biology. Nobel laureate Frances Arnold (@francesarnold), Linus Pauling Professor, California Institute of Technology, cofounder of several synthetic biology startups, is a US west coast resource. Lastly, Ed Boyden (@eboyden3), whose approach to synthetic neurobiology and neurotechnology is breaking new ground in terms of being able to analyze biological systems such as the brain and with the near-term promise of repairing neural dysfunction through optogenetics, perhaps by showing you a movie that triggers the brain's immune system.

However, my advice would be to follow scientists who work in these distinguished people's labs rather than they themselves. The reason is that the next breakthrough is likely to come from their students, not from them, and also the students might be more active on social media and in sharing emerging findings. For example, an influencer to track would be Reshma Shetty (@reshmapshetty), cofounder at Ginkgo Bioworks (@Ginkgo).

There are a few community innovators that come to mind, too, such as John Cumbers (@johncumbers), who has a PhD in molecular biology from Brown University, founded SynBioBeta, publishes a weekly newsletter, hosts the company's podcast (The SynBioBeta Podcast), and wrote a book called *What's Your Biostrategy?*

The main institutes for synthetic biology are found around Boston (most notably at Harvard and MIT), in California (Stanford, UCSF, UC Berkeley, UCLA, UCSD, Salk Institute, Caltech), around Paris, and in the UK (Imperial College London, Cambridge University, Oxford University), and with a rising interest in China (The Chinese Academy of Sciences).

Companies active in the synthetic biology market include Amyris Inc, Thermo Fisher Scientific Inc., Synthetic Genomics Inc., ATG: Bayer AG, Bristol-Myers Squibb Co., Exxon Mobil Corp., Ginkgo Bioworks, and many more. A unique company to look out for is Synthego (@Synthego), which is a genome engineering platform company founded in 2012, meaning it enables broader access to CRISPR-based gene editing and is used by scientists from both industry and academia.

SynBioBeta (https://synbiobeta.com/) is the premier innovation network for biological engineers, investors, innovators, and entrepreneurs who share a passion for using biology to build "a better, more sustainable universe." Based at Imperial College London, SynbiCITE (http://www.synbicite.com/)

is the Innovation and Knowledge Centre (IKC) for Synthetic Biology. Bio-start, the UK's first synthetic biology accelerator program, launched in 2017, is sponsored by SynbiCITE, Rainbow Seed Fund, Cambridge Consultants, Imperial Innovations, Analytik-Jena, BioCity, UnitDX Science Incubator, Norwich Research Park, Keltie, and Twist Biosciences.

Synbio writing

There is a plethora of specialty publications in synthetic biology. The best way to track the overall field is likely still to read the big journals, such as *Nature* and *Science*, each with a set of associated publications. PubMed (https://pubmed.ncbi.nlm.nih.gov/) is another great open access avenue, since journals are expensive unless you work somewhere that pays for such things. Most academic labs will have their newsletters that you could sub-scribe to free of charge. One general introduction is the book *Synthetic: How Life Got Made* (2017) by Sophia Roosth.

Synthetic biology conferences worth exploring

Unless you are in the field, there are only a few tech conferences you should attend where synthetic biology will be explored in any depth yet is ap-proachable (and even understandable). It is still a specialty field. The Global Synthetic Biology Summit (by SynBioBeta) might be the exception because despite being a conference, it is also a strong and supportive community.

Resources for 3D printing

3D printing people you should track (scientists, innovators, startup founders)

Yoav Zeif (https://www.linkedin.com/in/yoav-zeif-990117/) is the newly ap-pointed CEO of Stratasys and Jeff Graves (https://www.linkedin.com/in/jeffrey-graves-a0724b1/) is the newly appointed CEO of 3D Systems.

Natan Linder is co-founder of 3D printing pioneer MIT spinout Formlabs as well as co-founder of manufacturing app platform Tulip. Formlabs spun off from the MIT Media Lab in 2011 to develop a high-quality, low-cost desktop 3D printer. Formlabs runs the podcast The Digital Factory hosted by Jon Bruner. Another MIT-startup, Inkbit, founded in 2017 by Davide Marini (https://www.linkedin.com/in/dmarini/ and @davide_m_marini), provides multimaterial 3D printing.

The top 3D printing podcast is 3D Printing today by Andy Cohen and Whitney Potter, who have produced more than 358 weekly episodes on current events and technology, with guests ranging from specialists to luminaries. The Making Chips podcast for manufacturing leaders, so far with 400-plus episodes, covers the broader manufacturing business community, and Advanced Manufacturing Now from SME (https://www.sme.org/), a non-profit association of professionals, educators, and students committed to promoting and supporting the manufacturing industry, with 265-plus episodes, is a leading source for news. The best podcast search engine is ListenNotes (https://www.listennotes.com/) and you can also search for individual episodes.

See https://bind40.com/2020/03/24/industry40-influencers-to-follow.

3D printing writing

The top general publications for industry include Industry World (https://www.industryweek.com/), founded in 1882. An example of a decent specialty publication would be Metal AM (https://www.metal-am.com/).

Wohlers Report 2020, is the 25th annual installment of its industry-leading report on additive manufacturing (AM) and 3D printing. Often referred to as the "bible" of 3D printing, the new report provides a unique window into the constantly advancing global industry. The publication provides countless details on AM applications, software, workflows, materials, systems, and post-processing. It gives specifics on patents, startups, invest.

3D printing conferences

The manufacturing industry has no shortage of conferences but only a few of them are a must to track the progress of the industry. Additive Manufacturing Forum (AM Forum) since 2015 is the top European industry event as it brings together the whole value chain (https://am-forum.eu/). If you were to attend only one trade fair, the legendary Hannover Messe, which dates back to 1947 as Germany's first export fair, would be the one as it covers the industrial transformation value chain (from engineering, automation, energy, R&D, and industrial IT to logistics and subcontracting), with a focus on current buzzword topics such as Industry 4.0, digital platforms, AI, 5G, and hydrogen and fuel cells (https://www.hannovermesse.de/en/).

The International Manufacturing Technology Show (IMTS) is the biggest manufacturing technology trade show in the US, with roots in the National Machine Tool Builders' Exposition of 1927, presently covering more than 15,000 new machine tools, controls, computers, software, components, systems, and processes. At its height it has had over 100,000 attendees. The show is managed by the Association for Manufacturing Technology (https://www.amtonline.org/), which is another key source of industry insight. Beyond these, there are specific conferences for each geographical region (most countries or even big cities will have their own events), industry, and specific applications of 3D printing, such as metal 3D printing conferences. Also, each major corporate player in the space has its own event, e.g. Stratasys. Finally, for manufacturing there are material-specific events, e.g. The Plastics Show (https://npe.org/), produced by the Plastics Industry Association, founded in 1937, a global platform for industry knowledge and technological innovation with over 2,000 exhibiting companies; the plastics industry being a $432.3 billion industry according to the association itself (Plastics, 2018).

These emerging fields of course have their own taxonomies. For quick illustration's sake, let's take digital fabrication.

Digital fabrication

Oxide semiconductors

3D integration

Materials beyond silicon

High K dialectrics

Low K dielectrics

Plasmonics

Spintronics

Multiferroics

Each of those terms also has its own nomenclature. What we can clearly see from this type of analysis is that the specificity needed to operate as an efficient student of emerging science and technology fields is astounding. Needless to say, it rapidly becomes so complex that even recently graduated engineers in one of these subdisciplines could get lost, depending on how much time they spent reviewing the adjacent fields versus simply developing one narrow expertise.

Polymaths throughout history

Antiquity

Aristotle (384–322 BC), the Greek philosopher, was an expert in physics, metaphysics, poetry, theater, music, logic, rhetoric, politics, ethics, biology, and zoology. He was transdisciplinary and combined many of these topics in completely novel ways. Aristotle grew up and was taught in the Macedonian court in what today is called the Republic of North Macedonia, a country on the Balkan Peninsula in southeast Europe. In 367 B.C. at the age of 17, Aristotle, due to tensions at the court, went to Athens to study with Plato. Plato's Academy was the intellectual center of the Greek world and people traveled from all over to study, learn, and teach (Great Thinkers, 2020).

After Plato's death, Aristotle, in political danger (he was viewed as an alien aligned with the Macedonians), escaped Athens and began eight years of intellectually stimulating travels. In 335 B.C., Aristotle founded the Lyceum, his own school in Athens, where he spent most of the rest of his life studying, teaching, and writing. From this, I think we can learn that a combination of spending significant time with top mentors at the key learning sites plus educational travel can do wonders for your thinking skills.

Middle Ages

Hildegard of Bingen (1098–1179), a Benedictine nun rising to the stature of abbess, often described as a German mystic for her religious views of harmony through God's creation, had visions as a child, which were later endorsed by the archbishop of the important German city of Mainz as the "gift of prophecy" and which Pope Benedict XVI said have "rich theological content" (Franciscan Media, 2020). Hildegard was an artist, preacher, lyrical poet, and the best-known composer of sacred monophony. Yet, she was a scientific mind, herbalist, and early "doctor" who established the foundations of female health and anatomy. She offered dietary guidelines as treatment for amenorrhea in one of her books. She was the female scientific pioneer of the Middle Ages (Exploring Your Mind, 2020), yet was canonized by the Catholic Church after her death (Britannica, 2020b), which are both quite remarkable. In her text on science and medicine, *Causae et Curae*, she writes, "One ought to eat at the opportune time, before night falls, so that one can take a walk before lying down to sleep" (page 86), in an effort to instill good habits, which I try to follow myself.

The Renaissance

Ibn Khaldun (1332–1406) was a brilliant North African polymath of Arab descent who was born in Tunis but travelled extensively throughout North Africa. He was a statesman, philosopher, Islamic theologian and jurist, historian, astronomer, mathematician, economist, poet, and social scientist, and is widely considered to be the father of historiography, cultural history, demography, philosophy of history, and sociology.

Khaldun's view of learning was as a path toward self-understanding which would enable individuals to acquire mental thinking processes, such as comprehension, coherence, and cohesion through cognitive maturity that involved knowledge unity (away from the splitting up that modernization's fast rise in knowledge contributed to). He insisted on not moving on from one scientific question to another before the first was fully and deeply understood, and insisted on a multidisciplinary approach but always with one subject in the foreground, in order to avoid confusion (Halabi, 2013).

Khaldun saw learning itself as the acquisition of a "habitus" (malaka), which he defined as "a stable quality resulting from a *repeated action until its form has taken final shape.*" The choice of content in the earliest instruction is of decisive importance because the first teachings become imprints. Habitus appears to be an Aristotelian concept appropriated and actively used by the Arab thinkers (including Avicenna) and then by French sociologist Bourdieu much later. Either way, Khaldun's view was that habitus is a corporal phenomenon, something the soul can only gradually acquire through the senses but that stays in/on the body "like dye on a cloth" once it's applied. From this, we can take in that the process of acquiring expertise should ideally begin early (Cheddadi, 2009).

Leonardo da Vinci (1452–1519) was of course an Italian Renaissance painter, inventor, engineer, astronomer, anatomist, biologist, geologist, physicist, and architect (Capra, 2013). Leonardo is why we have the (somewhat sexist by today's standards) phrase "Renaissance man." No matter the label, he defied conventions and disciplines, and reframed entire epistemologies. He also had an immense capability for concentrating his efforts, even as he became involved in an enormous number of different explorations. The seven da Vincian principles—curiosity, testing through experience (the empirical method), refinement of the senses, willingness to embrace ambiguity, balancing art and science (depict rather than describe), cultivating both fitness and poise, and systems thinking—are timeless reminders of the potential for learning and creativity necessary to make use of technology or any other complex endeavor (Capra, 2013; Gelb, 2000).

Late modern period

Madame de Staël (1766–1817), a key European precursor to feminism, was a woman of letters (of travel literature, novels, and polemics) and a political theorist advocating constitutionalism and representative government, and a mega-networker running famous literary and political salons (not the hairdresser kind), who influenced European politics, brought German philosophy to France, and influenced Romanticism. She both understood and engaged with the events of the 19th century to an extremely influential degree, interacting with poets (Lord Byron in England and the Von Schlegel brothers in Germany), intellectual activists (Benjamin Constant), and politicians (Napoleon). She also wrote the first treatise on the sociology of literature. From her, we can learn the importance of being both *in the know* and *in the flow* in order to think and act both esthetically and politically.

Marie Curie (1867–1934), a prodigy in both literature and math, was a Polish and French physicist and chemist who conducted pioneering research on radioactivity. She was the first woman in France to obtain her PhD in physics (reviewers said it was the greatest single contribution to science ever written), the first woman to win a Nobel Prize, the first person and only woman to win the Nobel Prize twice, and the only person so far to win the Nobel Prize in two different scientific fields. From those facts, we can learn that there is a value to being first, because you see things with new eyes. Her marriage to peer scientist Pierre Curie sealed their scientific destinies for both of them, which is another testament to great teamwork. However, it was Marie's boldness that separated her out: Curie's hypothesis about the subatomic particles released by uranium would revise the scientific understanding of matter at its most elemental level (Des Jardins, 2011). At the time, atoms were thought of as the smallest elementary particle.

Ludwig Wittgenstein (1889–1951) was an Austrian-British philosopher, logician, mathematician, architect, aeronautical engineer, and musician. His family home was a center of Viennese cultural life during one of its most dynamic phases. Many of the great writers, artists, and intellectuals of fin de siècle Vienna were regular visitors to the Wittgensteins' home. One can wonder to what degree early exposure to creative genius influenced him, which is perhaps answered by this one quote from his notebooks (Journal Entry, 12 October 1916): "What cannot be imagined cannot even be talked about."

Wittgenstein was one of the greatest philosophers of the 20th century. He inspired logical positivism (his early work) and the philosophy of language (his later work) as well as having deep reflections on psychology and ethics.

Gregory Bateson (1904–1980) was a British anthropologist, sociologist, semiotician, linguist, and cyberneticist. His theories transcend topics and reframe entire disciplines. In one of his books, *Mind and Nature* (1979:68), he makes the strikingly simple observation that it "takes two somethings to make a difference." He also points out that bringing the two together has value. He cites the brilliant early computer science thinker Von Neumann, who once remarked that for "self-replication among machines, it would be a necessary condition that two machines should act in collaboration". Bateson, and his followers, then applied this, too. For example, his thinking formed the foundation of today's burgeoning field of biosemiotics, which combines biology and semiotics (Harries-Jones, 2016).

Nikola Tesla (1856–1943), the Serbian-American inventor (of the alternating electric current among many other things), electrical engineer, mechanical engineer, theoretical and experimental physicist, mathematician, futurist, and humanitarian, is among the fathers of modern engineering. He obtained 300 patents from his activity. Tesla spoke eight languages and had a photographic memory, capable of memorizing entire books, and had, by his own estimation, extraordinary sight and hearing, "at least thirteen times more sensitive" than the ears of his assistants.

In his autobiography, he mentions how much he appreciates the "inestimable value of introspection" and describes his method as such: "When I get an idea I start at once building it up in my imagination. I change the construction, make improvements and operate the device in my mind. It is absolutely immaterial to me whether I run my turbine in thought or test it in my shop." Another poignant quote is the following: "On one occasion I started to read the works of Voltaire when I learned, to my dismay, that there were close on one hundred large volumes in small print which that monster had written while drinking seventy-two cups of black coffee per diem. It had to be done, but when I laid aside the last book I was very glad, and said, 'Never more!'" (Tesla, 1919).

Nobel Prize winner in biology Francis Crick (1916–2004) credited his background in physics for his discovery of the structure of DNA—a problem previously deemed unsolvable by modern biologists. Frustrated with physics and with no true merits to his name, in 1947 (a year he divorced from his first wife), at age 31, he switched careers to biology, applying the "gossip test" (the idea that the topics you gossip about must be the ones you care about), having to learn it from scratch. However, it might have been his long-standing friendship with colleague J. D. Watson that ensured they cracked the code together. In fact, Crick had a remarkable ability to form long-standing productive friendships with other scientists (NIH, 2020).

Richard Feynman (1918–1988), also a Nobel Prize winner, generated his ideas about quantum electrodynamics while watching a guy spin a plate on his fingers in a cafeteria (Tank, 2020). The Feynman technique included a notebook entitled "Things I Don't Know about" because the first thing to do is to identify that which you do not know (but want to figure out). The second step is to make sure you learn it so well that you can teach it to a child, and in the process of failing to do so, reveal and uncover your knowledge gaps (Gleick, 1992).

My personal insight ecosystem

Rather than discussing insight in the abstract, I thought I would share what my own insight ecosystem looks like as an example. The way I organize it will be in terms of importance and classified into the forces of disruption framework.

Right now, I have the following topics that I track based on being an expert in those topics and wanting to remain at the forefront (innovation, technology, startups, investment, standardization, leadership, strategy, sociology, global issues, sustainability) as well as wanting to build emerging expertise in the topic (podcasting, audio editing, video editing). Within technology, I am particularly avidly tracking AI, robotics, synthetic biology, industrial tech, and media tech.

For me, the primary source of insight over the past few decades has been books. I have consumed thousands of them, usually at least one per day. This practice started when I was a teenager and has continued to this point. I read all the books in my local library that were age appropriate (and some that were not)—this would have to amount to at least 5,000 books before I was 20. What did this yield? An acute awareness of a lot of cultural and political phenomena, a background in literature from which I know a lot of cultural references.

Nowadays, my primary source of insight on a daily basis is a combination of looking through online search engines using a variety of keywords within my various interests (which are far and wide) as well as getting email newsletters. I subscribe to about 20 daily or weekly newsletters. They mostly tend to be duplicative, so I spend very little time scanning through the content and seldom click my way into actual articles unless there is a controversial story and I know I have the best source at hand. Daily newsletters include Quartz, Nuzzel, The Boston Globe, Chicago Tribune, The Block (crypto), Wired, Crunchbase News, The Washington Post, South China Morning Post, Morning Brew, MAGNiTT Daily, PitchBook News, Good Morning from CNN, The New York Times, Inc42 Media (India), CB Insights (founder Anand Sanwal's newsletter), The Block, VA News, Time, MIT Technology Review, Utility Dive, New York Review Books, Small Business Administration, Vanity Fair, BostInno, Condé Nast Spotlight, EBAN (European Business Angels), This is Classical Guitar, Becker's Health IT and CIO Review, Inside Higher Ed, EU-startups (Thomas Ohr's newsletter), theSkimm, The Capitals (Euractiv's cross-EU newsletter, Sifted, World Economic Forum The Agenda Weekly, Policy People, Foreign Affairs This Week, AngelList Weekly, Rocky Mountain Institute, Inside AI, Briefings from Goldman Sachs, Forbes | Under 3, AI2 Newsletter (Allen Institute for AI), ProFellow Insider, Wine Scholar Guild, Morning Report (by entrepreneurs for entrepreneurs), Benzinga (financial news), Search Engine Watch, NENSA (New England Nordic Ski Association), Inside AI, FS (Farnam Street), LinkedIn News, TrustedInsight Weekly News, LoveReading, GeoTech Center Atlantic Council, German-American Business Council, Society of Authors, GE Brief, and The Old Farmer's Almanac. I try to check at least one newspaper from each part of the world every day.

For technology, there is a plethora of online magazines with great track record (*Wired*, etc.) that I read regularly.

For industries, I track the main trade publications in the verticals I care about and I often read a bit wider than that.

For government policy, I track EU developments through various official EU websites (the European Commission, European Parliament), plus the G7 government websites. I don't have a regular schedule for this tracking but I'm trying to become more structured about it and will do so with my podcasting efforts.

As for newspapers, I read the *South China Morning Post* (China), *The Washington Post* (USA), *France 24* (France), *The Guardian* (UK), *El Pais* (Spain), *La Repubblica* (Italy), as well as *VG* (Norway). I mostly read newspapers that are free of charge, which means that I don't really read all the stories from *The New York Times*, the *FT*, *The Washington Post*, *Le Monde* (France) that I would like to read—I just don't consider the subscription price worth it with so many free sources available.

Education TV and news

When it comes to TV, I do very little. Historically, I've never watched TV. In fact, there were years when my family did not own a TV. When I grew up, we were among the last to get a color TV in my neighborhood, so much so that the license police came to our door and wanted to come into our house to verify the situation. I occasionally watch CNN and do a one-minute check with Fox News just to compare, but I find both largely biased and redundant. The talking heads keep repeating their guests and rant about the same issues, day out and day in. CNN is no longer first with news, that's typically a news alert directly from a country source.

Events I attend

Before COVID-19, I would attend one or two events in each interest area per year, meaning I would attend two corporate venture conferences (GCV Summit and GCV ...), one private equity conference (SuperReturn International), one venture capital event (SuperVenture), one founder event (Founders Forum), and one local innovation event (the Nantucket Conference). My sense is that this was too much, unless there was a compelling reason to meet clients or prospective investors specifically.

Consultants I use

I would not pay a dime for consulting, apart from highly specific individual service providers for software programming or marketing/PR, using either my own platforms to find them or a platform such as Fiverr.

Partnerships I rely on

This has varied throughout my career. I am typically in favor of partnerships as long as it is clearly spelled out what the deal is. Good partners can teach you a lot. When I was at Oracle, I partnered with IBM as well as a host of trade associations. When I was at MIT, I partnered with all the labs at the Institute.

Emerging insight ecosystems

What I would love to have is a more systematic way to scan through information and ideally have an avatar or an assistant pre-read information and present it to me either in semidigested format or as a prioritized, labeled reading list, contextualized with why the information is relevant. In fact, I used to take part in writing briefings to EU commissioners. They would get a 20-page dossier a few days before every meeting. Wouldn't that be sweet? We need an equivalent that can be accessible to every professional. Right now, it consists of looking up people's LinkedIn page and putting the meeting topic into a search engine half an hour before a meeting. This is clearly not optimal.

Industry insight ecosystems

There are many insights ecosystems for business functions like retail, micro-marketing, human capital management, supply chain optimization, delivery optimization as well as in every industry: banking, automotive, media, manufacturing, and so on (Hopkins and Schadler, 2018). Some of those ecosystems are provided by market research agencies and conglomerates (WPP, Omnicom, Ipsos, Publicis Groupe, Interpublic Group, Nielsen, and Dentsu).

Having something to offer

A big principle to follow in terms of gaining insight is to make sure you have something to offer yourself. The offer can be insight, or it can be access, or both. Being a recognized specialist, author, or consultant is the typical way to do this. Another is to become a media yourself.

Right now, I am building the podcast Futurized. Futurized goes beneath the trends, tracking the underlying forces of disruption in tech, policy, business models, social dynamics, and the environment. I have had numerous experts on the show already and the requests keep flowing in, for which I am grateful.

The advantage of becoming media yourself is that (a) I now have a valuable media spot to offer (being interviewed) and (b) I get to learn from all the topics and people I have on the show. This suddenly puts me in the leading seat both in terms of being *in the know* and being *in the flow*.

BIBLIOGRAPHY

Aguilar, F.J. (1967) *Scanning the Business Environment*, Macmillan, New York.

AI Multiple (2020) Top 61 RPA usecases/applications/examples in 2020, AI Multiple, 4 July. Available from: https://research.aimultiple.com/robotic-process-automation-use-cases/ (archived at https://perma.cc/GY36-7QPF) [Last accessed 7 July 2020]

AlterEgo (2020) AlterEgo Project, MIT Media Lab. Available from: www.media.mit.edu/projects/alterego/overview/ (archived at https://perma.cc/7NSA-4XDA) [Last accessed 26 June 2020]

Anderson, K. (2013) Whoops! Are some current open access mandates backfiring on the intended beneficiaries? The Scholarly Kitchen, Official Blog of the Society for Scholarly Publishing, 12 March. Available from: https://scholarlykitchen.sspnet.org/2013/03/12/whoops-are-some-current-open-access-mandates-backfiring-on-the-intended-beneficiaries/ (archived at https://perma.cc/25C7-YRBR) [Last accessed 21 May 2020]

ARENA (2014) Technology Readiness Levels for Renewable Energy Sectors, Australian Renewable Energy Agency. Available from: https://arena.gov.au/assets/2014/02/Technology-Readiness-Levels.pdf (archived at https://perma.cc/G3DC-VYHG) [Last accessed 4 May 2020]

Aridi, A. and Urška Petrovčič, U. (2020) How to regulate big tech, 13 February. Available from: www.brookings.edu/blog/future-development/2020/02/13/how-to-regulate-big-tech/ (archived at https://perma.cc/2KY6-TH2V) [Last accessed 9 June 2020]

Armstrong, B. (2020) What will happen to cryptocurrency in the 2020s, Coinbase, 3 January. Available from: https://blog.coinbase.com/what-will-happen-to-cryptocurrency-in-the-2020s-d93746744a8f (archived at https://perma.cc/36R4-D7XB) [Last accessed 10 July 2020]

Associated Press (2020) Segway, personal vehicle known for high-profile crashes, ending production. Available from: www.theguardian.com/technology/2020/jun/23/segway-transporter-production-ends (archived at https://perma.cc/HBB2-GJ45) [Last accessed 26 June 2020]

Atlantic Council (2020) Why data trusts could help us better respond and rebuild from COVID-19 globally, 15 April. Available from: www.atlanticcouncil.org/event/why-data-trusts-could-help-us-better-respond-and-rebuild-from-covid19-globally/ (archived at https://perma.cc/D3XU-NPPL) [Last accessed 12 September 2020]

Augsburg, T. (2014) Becoming transdisciplinary: the emergence of the transdisciplinary individual, world futures, *The Journal of New Paradigm Research*, 70, 3–4, 233–247. Available from: http://dx.doi.org/10.1080/0260402 7.2014.934639 (archived at https://perma.cc/RH5P-PVH5) [Last accessed 25 July 2020]

Baron, J. and Pohlmann, T. (2013) Who cooperates in standards consortia – rivals or complementors? *Journal of Competition Law & Economics*, 9, 4, December, 905–929. Available from: https://doi.org/10.1093/joclec/nht034 (archived at https://perma.cc/3FNR-X83R) (also see free full text earlier working paper version: www.law.northwestern.edu/research-faculty/clbe/workingpapers/ documents/Baron_Pohlmann_2013_Who_cooperates_in_standards_consortia% 20.pdf (archived at https://perma.cc/6WCD-3FGN)) [Last accessed 8 June 2020]

Baxter, R.K. (2020) *The Forever Transaction*. McGraw-Hill, New York.

BBC Science Focus (2019) Everything you need to know about Neuralink, BBC Science Focus, 9 October. Available from: www.sciencefocus.com/future-technology/everything-you-need-to-know-about-neuralink/ (archived at https:// perma.cc/9XHN-N4LS) Last accessed 7 August 2020]

Bell, J. (2020) Will hospitals use 3D printing to take greater ownership of supply chains after Covid-19 crisis? NS Medical Devices, 29 May. Available from: www.nsmedicaldevices.com/analysis/3d-printing-covid-19-crisis/ (archived at https://perma.cc/7HHG-WQQ2) [Last accessed 12 July 2020]

Bellmund, J. (2020) New preprint on bioRxiv: learning to navigate abstract knowledge, Doeller Lab, 18 July. Available from: https://doellerlab.com/ new-preprint-on-biorxiv-learning-to-navigate-abstract-knowledge/ (archived at https://perma.cc/WA6P-RMKE) [Last accessed 7 August 2020]

Bernard, A. (2020) Top 10 tech policy trends to watch in 2020, Tech Republic, 30 January. Available from: www.techrepublic.com/article/top-10-tech-policy-trends-to-watch-in-2020/ (archived at https://perma.cc/B69S-DRQG) [Last accessed 10 June 2020]

Bijker, W.E., Hughes, T.P. and Pinch, T.J., eds. (1987) *The Social Construction of Technological Systems: New Directions in the Sociology and History of Technology*. MIT Press, Cambridge, MA.

Bio (2020) Current uses of synthetic biology, BIO.org. Available from: https:// archive.bio.org/articles/current-uses-synthetic-biology# (archived at https:// perma.cc/7CDQ-W87H) [Last accessed 9 July, 2020]

Biography.com (2020) Mae C. Jemison. Available from: www.biography.com/ astronaut/mae-c-jemison (archived at https://perma.cc/3L7K-3RTE) [Last accessed 25 July 2020]

Biomechatronics (2020) Hugh Herr: Biomechatronics, MIT Media Lab. Available from: www.media.mit.edu/people/hherr/projects/ (archived at https://perma.cc/ 7QD2-MRXU) [Last accessed 7 August 2020]

Bizzabo (2019) 2020 Blockchain events: the #1 guide to blockchain conferences, Bizzabo Blog, 13 December. Available from: https://blog.bizzabo.com/blockchain-events (archived at https://perma.cc/AMP9-XFRD) [Last accessed 10 July 2020]

Blue Ocean Strategy (2020) Book website. Available from: www.blueoceanstrategy.com/ (archived at https://perma.cc/A48T-2WPT)

Board of Innovation (2020) 50+ business model examples, Boardofinnovation.com. Available from: www.boardofinnovation.com/guides/50-business-model-examples/ (archived at https://perma.cc/PY74-RVAL) [Last accessed 15 June 2020]

Boilard, M. (2018) Six ways technology is changing engineering, *Industry Week*, 23 October. Available from: www.industryweek.com/leadership/article/22026559/six-ways-technology-is-changing-engineering (archived at https://perma.cc/U3HF-7MFD) [Last accessed 14 August 2020]

Boissonneault, T. (2018) 10 ways 3D printing is positively impacting the world, 8 August. Available from: www.3dprintingmedia.network/ways-3d-printing-impacting-world/ (archived at https://perma.cc/X2XR-E2DC) [Last accessed 16 August 2020]

Bostrom, N. (2014) *Superintelligence*. Oxford University Press, Oxford.

Bouganim, R. (2014) What Is Govtech? Available from: http://govtechfund.com/2014/09/what-is-govtech/ (archived at https://perma.cc/ENK9-KMWW) [Last accessed 12 May 2020]

Bourdieu, P. (1977) *Outline of a Theory of Practice*. Cambridge University Press, Cambridge.

Boyle, P. (2019) Microbes and manufacturing: Moore's law meets biology, *The Bridge*, 49, 4. Available from: www.nae.edu/221231/Microbes-and-Manufacturing-Moores-Law-Meets-Biology (archived at https://perma.cc/2XVQ-REM4) [Last accessed 8 May 2020]

BRAIN Initiative (2020) Available from: https://braininitiative.nih.gov/ (archived at https://perma.cc/B6N6-7ZRU) [Last accessed 7 August 2020]

Brandom, R. (2019) Facebook's Libra Association crumbling as Visa, Mastercard, Stripe and others exit, 11 October. Available from: www.theverge.com/2019/10/11/20910330/mastercard-stripe-ebay-facebook-libra-association-withdrawal-cryptocurrency (archived at https://perma.cc/KKZ3-AAEP) [Last accessed 10 July 2020]

Brown, T. (2009) *Change by Design: How Design Thinking Transforms Organizations and Inspires Innovation*. HarperBusiness, New York.

Brownell, L. (2020) The next decade in science. Wyss Institute, Harvard. Available from: https://wyss.harvard.edu/news/the-next-decade-in-science/ (archived at https://perma.cc/RE5K-FUCS) [Last accessed 9 July 2020]

Brownsword, R. and Yeung, K., eds. (2008) *Regulating Technologies*. Hart Publishing, Oxford.

Britannica (2020a) Julie Taymor. Available from: www.britannica.com/biography/Julie-Taymor (archived at https://perma.cc/RSY3-NBXK) [Last accessed 25 July 2020]

Britannica (2020b) St. Hildegard. Available from: www.britannica.com/biography/Saint-Hildegard (archived at https://perma.cc/WLD5-PHW8) [Last accessed 25 July 2020]

Britzky, H. (2019) The Army wants to stick cyborg implants into soldiers by 2050 and it's absolutely insane, Task & Purpose, 27 November. Available from: https://taskandpurpose.com/military-tech/army-cyborg-soldier-2050-study# (archived at https://perma.cc/29L9-ZVLM) [Last accessed 7 August 2020]

Bryant, J. (2019) How AI and machine learning are changing prosthetics, MedTechDive, 29 March. Available from: www.medtechdive.com/news/how-ai-and-machine-learning-are-changing-prosthetics/550788/ (archived at https://perma.cc/RL4S-KP5B) [Last accessed 7 August 2020]

Cambridge Consultants (2018) Building the business of biodesign: the synthetic biology industry is ready to change gear, Cambridge Consultants, Workshop Report. Available from: www.cambridgeconsultants.com/sites/default/files/uploaded-pdfs/Building%20the%20business%20of%20biodesign%20%28workshop%20report%29_0.pdf (archived at https://perma.cc/XH5E-5PPD) [Last accessed 20 August 2020]

Cameron, D., Bashor, C., and Collins, J. (2014) A brief history of synthetic biology, Nature Reviews Microbiology, 12, 381–390. Available from: https://doi.org/10.1038/nrmicro3239 (archived at https://perma.cc/E2TL-DBDT) and https://collinslab.mit.edu/files/nrm_cameron.pdf (archived at https://perma.cc/ABP3-9C9R) (full text) [Last accessed 9 July 2020]

Caplan, A.L. (2020) Top 10 biomedical issues of the next decade, Genetic Engineering News, 40, 2, 1 February. Available from: www.genengnews.com/commentary/point-of-view/top-10-biomedical-issues-of-the-next-decade/ (archived at https://perma.cc/7H32-HYJV) [Last accessed 8 July 2020]

Capps, M. (2015) Nashville's Roger Brown sells Segway, but continues M&A, Venture Nashville, 15 April. Available from: www.venturenashville.com/nashvilles-roger-brown-sells-segway-but-continues-m-a-cms-1079 (archived at https://perma.cc/9PQV-DFDX) [Last accessed 26 June 2020]

Capra, F. (2013) Learning from Leonardo, Berrett-Koehler, San Francisco.

Carey, J. and Elton, M.C.J. (2010) When media are new: understanding the dynamics of new media adoption and use, University of Michigan Digital Culture. Available from: http://dx.doi.org/10.3998/nmw.8859947.0001.001 (archived at https://perma.cc/UB3S-DVCQ) [Last accessed 26 June 2020]

Carnegie, L.P. et al. (2020) How final CFIUS regulations will impact technology companies and investors, Program on Corporate Compliance and Enforcement (PCCE), New York University School of Law, 4 March. Available from: https://wp.nyu.edu/compliance_enforcement/2020/03/04/how-final-cfius-regulations-will-impact-technology-companies-and-investors/ (archived at https://perma.cc/C2TJ-VYGH) [Last accessed 22 May 2020]

Chao, J. (2018) What's on your skin? Archaea, that's what, Berkeley Lab, 12 February 2018. Available from: https://newscenter.lbl.gov/2017/06/29/whats-on-your-skin-archaea-thats-what/ (archived at https://perma.cc/ND54-Z9TH) [Last accessed 8 April 2020]

Chapman, J. (2015) Specialization, polymaths and the Pareto principle in a convergence economy, TechCrunch. Available from: https://techcrunch.com/2015/10/17/specialization-polymaths-and-the-pareto-principle-in-a-convergence-economy/ (archived at https://perma.cc/PTD6-7NP9)

Cheddadi, A. (2009) Ibn Khaldun's concept of education in the 'Muqaddima', Muslim Heritage, 15 May. Available from: https://muslimheritage.com/ibn-khalduns-education-muqaddima/#sec_5 (archived at https://perma.cc/6A7H-DWUF) [Last accessed 25 July 2020]

Chen, A. (2020) The EU just released weakened guidelines for regulating artificial intelligence, MIT Technology Review, 19 February. Available from: www.technologyreview.com/2020/02/19/876455/european-union-artificial-intelligence-regulation-facial-recognition-privacy/ (archived at https://perma.cc/2RPB-UEGC) [Last accessed 22 May 2020]

Cherdo, L. (2020) Metal 3D printers in 2020: a comprehensive guide, 13 July. Available from: www.aniwaa.com/buyers-guide/3d-printers/best-metal-3d-printer/ (archived at https://perma.cc/MV2D-A3E7) [Last accessed 16 August 2020]

Chiappone, J. (2020) List of polymaths, JohnChappione.com. Available from: www.johnchiappone.com/polymaths.html (archived at https://perma.cc/9C3P-2CVT) [Last accessed 25 July 2020]

Chiappone, J. (2020) Specialization, polymaths and the Pareto principle in a convergence economy, 17 October. Available from: https://techcrunch.com/2015/10/17/specialization-polymaths-and-the-pareto-principle-in-a-convergence-economy/ (archived at https://perma.cc/PTD6-7NP9) [Last accessed 25 July 2020]

ChiefMartec.com (2020) Chief Marketing Technologist Blog by Scott Brinker. Available from: https://chiefmartec.com/ (archived at https://perma.cc/F4RF-B5MH) [Last accessed 12 May 2020]

Choudhury, A. (2019) Top 8 Neuralink competitors everyone should track, Analytics India Magazine, 24 July. Available from: https://analyticsindiamag.com/top-8-neuralink-competitors-everyone-should-track/ (archived at https://perma.cc/GB5V-U664) [Last accessed 5 August 2020]

CIPD (2020) PESTLE analysis. Available from: www.cipd.co.uk/knowledge/strategy/organisational-development/pestle-analysis-factsheet#7994 (archived at https://perma.cc/NWH4-BJ3U) [Last accessed 13 August 2020]

Clarke, A.R. (2013) The end of an era, Journal of Neurotherapy: Investigations in Neuromodulation, Neurofeedback and Applied Neuroscience, 17, 4, 201–202. Available from: DOI: 10.1080/10874208.2013.855477 [Last accessed 10 August 2020]

CNET (2020) Neuralink: Elon Musk's entire brain chip presentation in 14 minutes (supercut), CNET, 28 August. Available from: www.youtube.com/watch?v= CLUWDLKAF1M (archived at https://perma.cc/B2UH-4HUF) [Last accessed 14 September 2020]

CNN (2018) Segway was one of the most hyped products of the century. Here's why it failed, CNN Business [video], 25 October 2018. Available from: www. cnn.com/videos/business/2018/10/25/what-happened-to-segway-orig.cnn-business (archived at https://perma.cc/G8DP-8JDP) [Last accessed 26 June 2020]

Cohen, R.M. (2020) To develop a Covid-19 vaccine, pharma and the federal government will have to break old patterns, The Intercept, 27 March. Available from: https://theintercept.com/2020/03/27/us-government-vaccines-big-pharma/ (archived at https://perma.cc/DP8L-JSQA) [Last accessed 21 May 2020]

Coinbase (2019) The 2019 leaders in crypto education, Coinbase, 28 August. Available from: https://blog.coinbase.com/highereducation-c4fb40ecbc0e (archived at https://perma.cc/8UUQ-Q8XB) [Last accessed 10 July 2020]

Consensys (2020) Blockchain use cases and applications by industry. Available from: https://consensys.net/blockchain-use-cases/ (archived at https://perma.cc/ 9VKW-JZZ4) [Last accessed 4 July 2020]

Cornell (2020) Cornell Notes. Available from: http://lsc.cornell.edu/study-skills/ cornell-note-taking-system/ (archived at https://perma.cc/DLP8-M2QF) [Last accessed 25 July 2020]

Cox, M. (2019) Pentagon report predicts rise of machine-enhanced super soldiers, Military.com, 2 December. Available from: www.military.com/daily-news/ 2019/12/02/pentagon-report-predicts-rise-machine-enhanced-super-soldiers.html (archived at https://perma.cc/TEB5-DBBX) [Last accessed 7 August 2020]

Craciun, L. (2020) The 30+ best blockchain conferences to attend in 2020, Capsblock, 23 September. Available from: https://capsblock.io/blog/best-blockchain-conferences/ (archived at https://perma.cc/6C7W-F6JD) [Last accessed 10 July 2020]

CRS (2020) The Committee on Foreign Investment in the United States (CFIUS). Available from: https://fas.org/sgp/crs/natsec/RL33388.pdf (archived at https:// perma.cc/JT3E-RTBU) [Last accessed 22 May 2020]

Cryptochainuni (2020) List of crypto and blockchain media publications in the world, Crypto Chain University. Available from: https://cryptochainuni.com/ publication/ (archived at https://perma.cc/B9TB-TR7P) [Last accessed 10 July 2020]

Cryptoweekly (2020) The 100 most influential people in crypto, Cryptoweekly. Available from: https://cryptoweekly.co/100/ (archived at https://perma.cc/ KP4S-5V8B) [Last accessed 10 July 2020]

Cusumano, M.A., Yoffie, D.B., and Gawer, A. (2020) The future of platforms. MIT Sloan Management Review, Spring, 11 February. Available from: https:// sloanreview.mit.edu/article/the-future-of-platforms/ (archived at https://perma.cc/ 8LBV-TT4K) [Last accessed 4 May 2020]

Dafoe, A. and Russell, S. (2016) Yes, we are worried about the existential risk of artificial intelligence, *MIT Technology Review*, 2 November. Available from: www.technologyreview.com/2016/11/02/156285/yes-we-are-worried-about-the-existential-risk-of-artificial-intelligence/ (archived at https://perma.cc/QWU8-XP4L) [Last accessed 10 July 2020]

Dale, B. (2020) Tezos co-founder turns to gaming with 'Hearthstone' competitor, Coindesk, 18 March. Available from: www.coindesk.com/tezos-co-founder-turns-to-gaming-with-hearthstone-competitor (archived at https://perma.cc/G7BV-8QDQ) [Last accessed 10 July 2020]

Damani, A. (2016) 9 Productivity tools used by successful consultants. Available from www.linkedin.com/pulse/9-productivity-tools-used-successful-consultants-anand-damani/ (archived at https://perma.cc/W4M8-N857) [Last accessed 26 February 2018]

Dartmouth (2005) Dartmouth Artificial Intelligence Conference: The Next Fifty Years, Dartmouth College. Available from: www.dartmouth.edu/~ai50/homepage.html (archived at https://perma.cc/2GBC-3Z4L) [Last accessed 4 May 2020]

De Bono, E. (1967) *The Use of Lateral Thinking*, Cape, London.

Deline, B., Thompson, J.R., Smith, N.S. et al. (2020) Evolution and development at the origin of a phylum. *Current Biology*, 30, 1–8. Available from: www.cell.com/current-biology/pdf/S0960-9822(20)30261-X.pdf (archived at https://perma.cc/29AQ-T8MF) [Last accessed 8 April 2020]

Deloitte (2015) Boost your venturing results through data driven sourcing, Inside magazine issue 10, October 2015, www2.deloitte.com/content/dam/Deloitte/lu/Documents/technology/lu-boost-venturing-results-data-driven-venture-102015.pdf (archived at https://perma.cc/X824-96U9) (accessed 3/5/2018).

Delos Santos, J.M. (2020) Top 10 best project management software & tools in 2020, ProjectManagement.com, 13 April. Available from: https://project-management.com/top-10-project-management-software/ (archived at https://perma.cc/L9YY-B6YL) [Last accessed 4 May 2020]

Des Jardins, J. (2011) Madame Curie's passion. Available from: www.smithsonianmag.com/history/madame-curies-passion-74183598/ (archived at https://perma.cc/6LG4-J3BB) [Last accessed 10 July 2020]

Desai, F. (2015) The evolution of fintech, *Forbes*, 13 December. Available from: www.forbes.com/sites/falgunidesai/2015/12/13/the-evolution-of-fintech/#77a48e5a7175 (archived at https://perma.cc/XAG4-4UU8) [Last accessed 21 May 2020]

Detailed (2020) The 50 best cryptocurrency blogs, Detailed.com. Available from: https://detailed.com/cryptocurrency-blogs/ (archived at https://perma.cc/J4M8-E3DC) [Last accessed 10 July 2020]

Dewey, J. (2019) Blockchain & cryptocurrency regulation, Global Legal Insights. Available from: www.acc.com/sites/default/files/resources/vl/membersonly/Article/1489775_1.pdf (archived at https://perma.cc/RCH7-HP8H)

Diamandis, P. (2020) 7 business models for the next decade, Book excerpt, Diamandis.com. Available from: www.diamandis.com/blog/7-business-models-for-2020s (archived at https://perma.cc/Q2VM-DTMK) [Last accessed 12 May 2020]

Dolan, S. (2020) How the laws & regulations affecting blockchain technology and cryptocurrencies, like Bitcoin, can impact its adoption, *Business Insider*, 3 March. Available from: www.businessinsider.com/blockchain-cryptocurrency-regulations-us-global (archived at https://perma.cc/TPL6-HJPC) [Last accessed 4 July 2020]

Dresler, M., Sandberg, A., Bublitz, C. et al. (2019) Hacking the brain: dimensions of cognitive enhancement, *ACS Chem Neurosci*, 10, 3, 1137–1148. Available from: doi: 10.1021/acschemneuro.8b00571 [Last accessed 7 August 2020]

Drew, L. (2019) The ethics of brain–computer interfaces, *Nature*, 24 July. Available from: www.nature.com/articles/d41586-019-02214-2 (archived at https://perma.cc/DSX9-JX7C) [Last accessed 7 August 2020]

As technologies that integrate the brain with computers become more complex, so too do the ethical issues that surround their use.

Dybå, T., Dingsøyr, T., and Moe, N. (2014) Agile project management. 10.1007/978-3-642-55035-5_11. Available from: www.researchgate.net/publication/263276642_Agile_Project_Management (archived at https://perma.cc/376L-VC37) [Last accessed 5 May 2020]

EBSCO (2009) Four stages of social movements, EBSCO Research Starters: Academic Topic Overviews. Available from: www.ebscohost.com/uploads/imported/thisTopic-dbTopic-1248.pdf (archived at https://perma.cc/62BK-ZKU5) [Last accessed 27 June 2020]

EC (2020) Blockchain technologies, European Commission. https://ec.europa.eu/digital-single-market/en/blockchain-technologies (archived at https://perma.cc/NS22-GG39) [Last accessed 7 July 2020]

EC (2020a) White Paper: On Artificial Intelligence – A European approach to excellence and trust, European Commission. COM(2020) 65 final, 19 February. Available from: https://ec.europa.eu/info/sites/info/files/commission-white-paper-artificial-intelligence-feb2020_en.pdf (archived at https://perma.cc/Z57E-FZ5K) [Last accessed 22 May 2020]

EC (2020b) Guidelines on stakeholder consultation, European Commission. Available from: https://ec.europa.eu/info/sites/info/files/better-regulation-guidelines-stakeholder-consultation.pdf (archived at https://perma.cc/RYQ2-66P7) [Last accessed 11 June 2020]

EC (2020c) DG Competition>Competition>Cartels. European Commission. Available from: https://ec.europa.eu/competition/cartels/cases/cases.html (archived at https://perma.cc/ZDN6-AC9X) [Last accessed 11 June 2020]

Edwards, D. (2020) Amazon now has 200,000 robots working in its warehouses. Available from: XXX [Last accessed 7 July 2020]

Eggers, W.D., Turley, M., and Kishnani, P. (2018) The regulator's new toolkit. Technology and tactics for tomorrow's regulator. Available from: www2. deloitte.com/content/dam/Deloitte/br/Documents/public-sector/Regulator-4-0. pdf (archived at https://perma.cc/KNT3-GC5Y) [Last accessed 11 June 2020]

eLearning Learning (2020) Elearninglearning.com Available from: www. elearninglearning.com/ (archived at https://perma.cc/7D3G-LQBV) [Last accessed 21 July 2020]

Engler, A. (2020) The European Commission considers new regulations and enforcement for "high-risk" AI, TechTank Blog, Brookings Institution, 26 February. Available from: www.brookings.edu/blog/techtank/2020/02/26/ the-european-commission-considers-new-regulations-and-enforcement-for-high-risk-ai/ (archived at https://perma.cc/GAN8-J79C) [Last accessed 22 May 2020]

Ericsson, K.A., Krampe, R.T., and Tesch-Romer, C. (1993) The role of deliberate practice in the acquisition of expert performance, *Psychological Review*, 100, 3, 363–406. Available from: https://graphics8.nytimes.com/images/blogs/ freakonomics/pdf/DeliberatePractice(PsychologicalReview).pdf (archived at https://perma.cc/9CFA-WS3B) [Last accessed 14 September 2020]

ET Bureau (2020) Locus Robotics passes 100 million units picked for global retail and 3pl partners, breaking warehouse AMR robotics industry records, Enterprise Talk, 10 February. Available from: https://enterprisetalk.com/news/ locus-robotics-passes-100-million-units-picked-for-global-retail-and-3pl-partners-breaking-warehouse-amr-robotics-industry-records/ (archived at https://perma.cc/MYR9-P9BJ) [Last accessed 12 July 2020]

EU (2018) Did you know? EU funded research is shaping your future, EU Publication Office, 6 February. Available from: https://ec.europa.eu/ programmes/horizon2020/en/news/new-booklet-shows-how-eu-research-and-innovation-funding-impacts-your-daily-life (archived at https://perma.cc/ L585-3LFD) [Last accessed 21 May 2020]

EU (2019) European Parliament resolution of 12 February 2019 on a comprehensive European industrial policy on artificial intelligence and robotics (2018/2088(INI)), 12 February. Available from: www.europarl.europa.eu/doceo/ document/TA-8-2019-0081_EN.html (archived at https://perma.cc/6EJ9-7CYQ) [Last accessed 12 July 2020]

EU Parliament News (2020) EU action: research on Covid-19 vaccines and cures, EU Parliament News, 15 May. Available from: www.europarl.europa.eu/news/ en/headlines/society/20200323STO75619/eu-action-research-on-covid-19-vaccines-and-cures (archived at https://perma.cc/DR9W-MSC9) [Last accessed 21 May 2020]

Exploring Your Mind (2020) Hildegard of Bingen: biography of a female polymath, 21 July, Exploring Your Mind. Available from: https://exploringy-ourmind.com/hildegard-of-bingen-biography-female-polymath/ (archived at https://perma.cc/H5QP-SU8Q) [Last accessed 25 July 2020]

Eyal, N. (2016) 3 pillars of the most successful tech products, Techcrunch, 19 October. Available from: https://techcrunch.com/2016/10/19/3-pillars-of-the-most-successful-tech-products/ (archived at https://perma.cc/6HLK-MLNZ) [Last accessed 26 June 2020]

FDA (2019) Implanted brain-computer interface (BCI) devices for patients with paralysis or amputation – non-clinical testing and clinical considerations, February. Available from: www.fda.gov/regulatory-information/search-fda-guidance-documents/implanted-brain-computer-interface-bci-devices-patients-paralysis-or-amputation-non-clinical-testing (archived at https://perma.cc/2TUQ-HL5V) [Last accessed 7 August 2020]

Ferry, A. (2018) The evolution of retail tech: what we have learned, where we are and where we're headed, Retail TouchPoints, 11 September. Available from: https://retailtouchpoints.com/features/executive-viewpoints/the-evolution-of-retail-tech-what-we-have-learned-where-we-are-and-where-we-re-headed (archived at https://perma.cc/D354-MT5G) [Last accessed 12 May 2020]

Firestein, S. (2016) Failure: why science is so successful, Oxford University Press, Oxford.

Forrest, C. (2015) Tech nostalgia: the top 10 innovations of the 2000s, TechRepublic, 1 May. Available from: www.techrepublic.com/pictures/tech-nostalgia-the-top-10-innovations-of-the-2000s/ (archived at https://perma.cc/VZ5T-ZVAV) [Last accessed 13 August 2020]

Forrester, J.W. (1961) Industrial Dynamics, The MIT Press, Cambridge, MA. Reprinted by Pegasus Communications, Waltham, MA.

Fox, S. (2009) Gallery: The top 10 failed NASA missions, Popular Science, 10 March. Available from: www.popsci.com/military-aviation-amp-space/article/2009-03/gallery-top-10-nasa-probe-failures/ (archived at https://perma.cc/X9GK-MWNP) [Last accessed 8 April 2020]

Franciscan Media (2020) Saint Hildegard of Bingen. Available from: www.franciscanmedia.org/saint-hildegard-of-bingen/ (archived at https://perma.cc/K3M5-YMJN) [Last accessed 25 July 2020]

Frankenfield, J. (2019) What you should know about RegTech, Investopedia, 27 April. Available from: www.investopedia.com/terms/r/regtech.asp (archived at https://perma.cc/7ZEK-7JPF) [Last accessed 12 May 2020]

Fukuyama, F. (1992) The End of History and the Last Man, Free Press, New York.

Fung, I. (2019) The present and future of foodtech investment opportunity, TechCrunch, 22 October. Available from: https://techcrunch.com/2019/10/22/the-foodtech-investment-opportunity-present-and-future/ (archived at https://perma.cc/TY93-HSNJ) [Last accessed 12 May 2020]

Fussell, S. (2019) Did body cameras backfire? Nextgov, 2 November. Available from: www.nextgov.com/emerging-tech/2019/11/did-body-cameras-backfire/161040/ (archived at https://perma.cc/2Y46-CE4Y) [Last accessed 20 May 2020]

GAO (2020) Technology readiness assessment guide: best practices for evaluating the readiness of technology for use in acquisition programs and projects, Government Accountability Office, 7 January. Available from: www.gao.gov/assets/710/703694.pdf (archived at https://perma.cc/H26E-F2ZQ) [Last accessed 4 May 2020]

Gartner (2020) Gartner Hype Cycle. Available from: www.gartner.com/en/research/methodologies/gartner-hype-cycle (archived at https://perma.cc/3WPP-R6KA) [Last accessed 12 September 2020]

Gassmann, O., Frankenberger, K., and Choudury, M. (2014) *The Business Model Navigator: 55 Models That Will Revolutionise Your Business*, FT Press, Upper Saddle River, NJ.

Gelb, M.J. (2000) *How to Think Like Leonardo da Vinci: Seven Steps to Genius Every Day*, Dell Publishing Company, New York.

Genomeweb (2020) Researchers rapidly reconstruct SARS-CoV-2 virus using synthetic genomics. Available from: www.genomeweb.com/synthetic-biology/researchers-rapidly-reconstruct-sars-cov-2-virus-using-synthetic-genomics#.XrG7w6hKiUk (archived at https://perma.cc/M9K7-JM8Y) [Last accessed 4 May 2020]

Gent, E. (2018) The 10 grand challenges facing robotics in the next decade, 6 February. Available from: https://singularityhub.com/2018/02/06/the-10-grand-challenges-facing-robotics-in-the-next-decade/ (archived at https://perma.cc/8XRK-K3GL) [Last accessed 6 July 2020]

Gilbert, N. (2020) 15 best team collaboration software reviews for 2020, Finances Online. Available from: https://financesonline.com/15-best-team-collaboration-software-reviews/ (archived at https://perma.cc/PGW5-N3TZ) [Last accessed 4 May 2020]

Gleick, J. (1992) *Genius: The Life and Science of Richard Feynman*, Vintage, New York.

Global Legal Insights (2020) Blockchain & cryptocurrency regulation 2020. 13 Legal issues surrounding the use of smart contracts, Gli Legal Insights. Available from: www.globallegalinsights.com/practice-areas/blockchain-laws-and-regulations/13-legal-issues-surrounding-the-use-of-smart-contracts (archived at https://perma.cc/J77E-4LRD) [Last accessed 7 July 2020]

Global Workplace Analytics (2020) Work-at-home after covid-19—our forecast, Global Workplace Analytics.com, Available from: https://globalworkplace-analytics.com/work-at-home-after-covid-19-our-forecast (archived at https://perma.cc/SJ7T-ENYL) [Last accessed 21 May 2020]

Globaldata (2019) Top ten blockchain influencers in Q3 2019, revealed by Globaldata. Available from: https://globaldata.com/top-ten-blockchain-influencers-in-q3-2019-revealed-by-globaldata/[Last (archived at https://perma.cc/TM5L-9CW3) accessed 10 July 2020]

Goldsberry, C. (2019) 3D printer on International Space Station allows astronauts to recycle, reuse, repeat, *Plastics Today*, 15 February. Available from: www.plasticstoday.com/3d-printing/3d-printer-on-international-space-station-allows-astronauts-recycle-reuse-repeat/97318583960275# (archived at https://perma.cc/SH3C-55KX) [Last accessed 7 July 2020]

Goode, L. and Calore, M. (2019) The 10 tech products that defined this decade, *Wired*, 25 December. Available from: www.wired.com/story/top-10-tech-products-of-the-decade/ (archived at https://perma.cc/4GQ2-KSGZ) [Last accessed 26 June 2020]

Goodfellow, I., Bengio, Y., and Courville, A. (2016) *Deep Learning*, MIT Press, Cambridge, MA.

GP-write (2020) Available from: https://engineeringbiologycenter.org/ (archived at https://perma.cc/R7EQ-FHU5) [Last accessed 5 May 2020]

Great Thinkers (2020) Great Thinkers: Aristotle: Biography. Available from: https://thegreatthinkers.org/aristotle/biography/ (archived at https://perma.cc/UGN5-TCGH) [Last accessed 25 July 2020]

Greenwood, B. (2014) The contribution of vaccination to global health: past, present and future. Philosophical Transactions of the Royal Society of London. Series B, Biological sciences, 369(1645), 20130433. Available from: https://doi.org/10.1098/rstb.2013.0433 (archived at https://perma.cc/WTL8-W4MT) [Last accessed 21 May 2020]

Gregurić, L. (2020) How much does a metal 3D printer cost in 2020? All3DP.com, 17 June. Available from: https://all3dp.com/2/how-much-does-a-metal-3d-printer-cost/ (archived at https://perma.cc/BRT7-UAS8) [Last accessed 16 August 2020]

Groves, H.T., Cuthbertson, L., James, P. et. al (2018) Respiratory disease following viral lung infection alters the murine gut microbiota. *Front. Immunol.*, 12 February 2018. Available from: https://doi.org/10.3389/fimmu.2018.00182 (archived at https://perma.cc/9ZSH-75FZ) [Last accessed 8 April 2020]

GSMA Intelligence (2020) GSMA Intelligence. Available from: www.gsmaintelligence.com/data (archived at https://perma.cc/MB3Y-GSJ6) [Last accessed 26 June 2020]

Gurría, A. (2020) Virtual 2020 G20 Digital Ministers Summit on COVID-19, 30 April. Available from: www.oecd.org/about/secretary-general/virtual-2020-g20-digital-ministers-summit-on-covid19-april-2020.htm (archived at https://perma.cc/BP52-9XK2) [Last accessed 9 June 2020]

Hackernoon (2020) Blockchain influencers in 2020, Hackernoon.com, 2 February. Available from: https://hackernoon.com/top-20-most-influential-people-in-blockchain-of-2020-xu2t33zq (archived at https://perma.cc/Y5BT-RKRC) [Last accessed, 10 July 2020]

Hague, P. (2019) *The Business Models Handbook: Templates, Theory and Case Studies*, Kogan Page, London.

Halabi, K.M.A. (2013) Ibn Khaldun's theory of knowledge and its educational implications, Institute of Education, International Islamic University Malaysia. Available from: https://lib.iium.edu.my/mom/services/mom/document/getFile/b4n7WiFqqprJQlrnRg1l0GZDFNtPF2AI20151022080814378 (archived at https://perma.cc/ZV32-4DXW) [Last accessed 12 September 2020]

Hanse, M.T. and Oetinger, B. von (2001) Introducing T-shaped managers: knowledge management's next generation, *Harvard Business Review*, March.

Harries-Jones, P. (2016) *Upside-Down Gods: Gregory Bateson's World of Difference*, Fordham University Press, New York.

Hartung, A. (2015) The reason why Google Glass, Amazon Fire Phone and Segway all failed, *Forbes*, 12 February. Available from: www.forbes.com/sites/adamhartung/2015/02/12/the-reason-why-google-glass-amazon-firephone-and-segway-all-failed/#7f37fa0bc05c (archived at https://perma.cc/6R4U-CBLT) [Last accessed 26 June 2020]

Hayden, E.C. (2017) The rise and fall and rise again of 23andMe: how Anne Wojcicki led her company from the brink of failure to scientific pre-eminence, *Nature News*, 11 October. Available from: www.nature.com/news/the-rise-and-fall-and-rise-again-of-23andme-1.22801 (archived at https://perma.cc/VFA5-LXAY) [Last accessed 5 June 2020]

Heilemann, J. (2001) Reinventing the wheel, *Time*, 2 December. Available from: http://content.time.com/time/business/article/0,8599,186660-5,00.html (archived at https://perma.cc/SS67-J33G) [Last accessed 9 June 2020]

Hernández-Ramos, P. (2000) Changing the way we learn: how Cisco Systems is doing it, IEEE Xplore, February. Available from: DOI: 10.1109/IWALT.2000.890601 [Last accessed 20 May 2020]

Hern, A. (2020) Volunteers create world's fastest supercomputer to combat coronavirus, *The Guardian*, 15 April. Available from: www.theguardian.com/technology/2020/apr/15/volunteers-create-worlds-fastest-supercomputer-to-combat-coronavirus (archived at https://perma.cc/4PJ2-3TLS) [Last accessed 4 May 2020]

HG.org (2020) What is the legal status of a Segway? Motor vehicle? Electric bike? Pedestrian? HG.org Legal Resources. Available from: www.hg.org/legal-articles/what-is-the-legal-status-of-a-segway-motor-vehicle-electric-bike-pedestrian-36294 (archived at https://perma.cc/8SAP-TUY3) [Last accessed 26 June 2020]

Hiley, C. (2020) Future mobile phones: what's coming our way? U Switch, 15 June. Available from: www.uswitch.com/mobiles/guides/future-of-mobile-phones/ (archived at https://perma.cc/PJ33-M479) [Last accessed 28 June 2020]

Hilton, M. (2007) Social activism in an age of consumption: the organized consumer movement, *Social History*, 32, 2, 121–143. Available from: www.jstor.org/stable/4287422 (archived at https://perma.cc/62ER-HKYU) [Last accessed 28 June 2020]

Holst, A. (2020) U.S. President's federal government IT budget 2015–2021, by department, Statista.com, 3 April. Available from: www.statista.com/statistics/605501/united-states-federal-it-budget/ (archived at https://perma.cc/Y25N-MRLM) [Last accessed 9 June 2020]

Hopkins, B. and Schadler, T. (2018) Digital insights are the new currency of business, 12 January. Available from: www.forrester.com/report/Digital+Insights+Are+The+New+Currency+Of+Business/-/E-RES119109 (archived at https://perma.cc/5T6T-W4VV) for $745 USD or by subscription [Last accessed 26 July 2020]

Hunt, T. (2019) Tam Hunt: Patient Zero — Liz Parrish talks longevity science, personal immortality, Opinions, 29 January. Available from: www.noozhawk.com/article/tam_hunt_liz_parrish_talks_longevity_science_personal_immortality_20190129 (archived at https://perma.cc/N9VX-JTRR) [Last accessed 7 August 2020]

Iansiti, M. and Lakhani, K.R. (2020) *Competing in the Age of AI: Strategy and Leadership When Algorithms and Networks Run the World*, Harvard Business Review Press, Cambridge, MA.

IBIS World (2020) The 10 fastest growing industries in the US. Available from: www.ibisworld.com/united-states/industry-trends/fastest-growing-industries/ (archived at https://perma.cc/R6KE-GN3T) [Last accessed 21 May 2020]

IEP (2020) Fallacy, Internet Encyclopedia of Philosophy. Available from: https://iep.utm.edu/fallacy/ (archived at https://perma.cc/LGY9-PGCT) [Last accessed 12 September 2020]

Institute of Medicine (1995) (US) Committee on Technological Innovation in Medicine; Rosenberg, N., Gelijns, A.C., Dawkins, H., eds. Sources of Medical Technology: Universities and Industry. Washington (DC): National Academies Press (US). 5, Cochlear Implantation: Establishing Clinical Feasibility, 1957–1982. Available from: www.ncbi.nlm.nih.gov/books/NBK232047/ (archived at https://perma.cc/2KAM-SLYK) [Last accessed 10 August 2020]

Insurance Information Institute (2020) Background on: Insurtech, 6 January. Available from: www.iii.org/article/background-on-insurtech (archived at https://perma.cc/2Y6E-MXCG) [Last accessed 12 May 2020]

Internet Society (2017) Internet society perspectives on internet content blocking: an overview, Internet Society. Available from: www.internetsociety.org/resources/doc/2017/internet-content-blocking/ (archived at https://perma.cc/A2HW-55K5) [Last accessed 12 May 2020]

IoM (2003) Institute of Medicine (US) Committee on the Evaluation of Vaccine Purchase Financing in the United States. Financing Vaccines in the 21st Century: Assuring Access and Availability. Washington (DC): National Academies Press (US); 2003. 2, Origins and Rationale of Immunization Policy. Available from: www.ncbi.nlm.nih.gov/books/NBK221822/ (archived at https://perma.cc/K9RJ-2G8D) [Last accessed 21 May 2020]

ISO (2020) Why ISO 9001? ISO.org. Available from: www.iso.org/iso-9001-quality-management.html (archived at https://perma.cc/FLL5-PTQ2) [Last accessed 8 June 2020]

Jemison, M. (2020) Bio available from: www.drmae.com/ (archived at https://perma.cc/6YLL-65BP) [Last accessed 25 July 2020]

Jeste D.V., Lee, E.E., Cassidy, C. et al. (2019) The new science of practical wisdom, *Perspect Biol Med.*, 62, 2, 216–236. Available from: doi:10.1353/pbm.2019.0011 [Last accessed 18 August 2020]

Jones, K. (2019) The Big Five: largest acquisitions by tech company, Visual Capitalist, 11 October. Available from: www.visualcapitalist.com/the-big-five-largest-acquisitions-by-tech-company/ (archived at https://perma.cc/7QS4-3XFY) [Last accessed 9 June 2020]

Joss, S. and Belucci, S., eds. (2002) *Participatory Technology Assessment: European Perspectives*. CSD, London.

Karn, M. (2019) Why the agtech boom isn't your typical tech disruption, World Economic Forum, 25 February. Available from: www.weforum.org/agenda/2019/02/why-the-agtech-boom-isn-t-your-typical-tech-disruption/ (archived at https://perma.cc/KYY8-C62S) [Last accessed 12 May 2020]

Joyce, S. (2013) A brief history of travel technology – from its evolution to looking at the future, PhocusWire, 12 September. Available from: www.phocuswire.com/A-brief-history-of-travel-technology-from-its-evolution-to-looking-at-the-future (archived at https://perma.cc/ZAV9-G25L) [Last accessed 12 May 2020]

Kemper, S. (2005) *Reinventing the Wheel: A Story of Genius, Innovation, and Grand Ambition*, HarperBusiness, New York.

Kharpal, A. (2020) Use of surveillance to fight coronavirus raises concerns about government power after pandemic ends, CNBC, March. Available from: www.cnbc.com/2020/03/27/coronavirus-surveillance-used-by-governments-to-fight-pandemic-privacy-concerns.html (archived at https://perma.cc/PJ2X-RDQX) [Last accessed 11 June 2020]

Kim, M. and Chun, J. (2014) New approaches to prokaryotic systematics. In *Methods in Microbiology*, 41, 2–327. Available from: www.sciencedirect.com/topics/neuroscience/16s-ribosomal-rna (archived at https://perma.cc/V4DS-CQBD) [Last accessed 8 April 2020]

Kim, W.C. and Mauborgne, R. (2005) *Blue Ocean Strategy: How to create Uncontested Market Space and Make the Competition Irrelevant*, Harvard Business School Press, Boston, MA.

Koch, R. (2018) Here are all the countries where the government is trying to ban VPNs, Proton VPN, 19 October. Available from: https://protonvpn.com/blog/are-vpns-illegal/ (archived at https://perma.cc/HFW6-QP3T) [Last accessed 14 August 2020]

Kolb, D.A. (1984) *Experiential Learning: Experience as the Source of Learning and Development*, Prentice-Hall, Englewood Cliffs, NJ.

Kolodny, L. (2017) Desktop Metal reveals how its 3D printers rapidly churn out metal objects, 25 April. Available from: https://techcrunch.com/2017/04/25/desktop-metal-reveals-how-its-3d-printers-rapidly-churn-out-metal-objects/ (archived at https://perma.cc/8A8W-YFE5) [Last accessed 16 August 2020]

Kraus, S. (2020) What is health tech and how will it continue to evolve? HotTopics.ht. Available from: www.hottopics.ht/23983/what-is-health-tech/ (archived at https://perma.cc/DS8K-D7Z2) [Last accessed 20 May 2020]

Krings, B., Rodríguez, H., and Schleisiek, A., eds. (2016) *Scientific Knowledge and the Transgression of Boundaries*, Springer VS, Wiesbaden.

Krishevsky, A., Sutskever, I., and Hinton, G.E. (2017) ImageNet classification with deep convolutional neural networks. Communications of the ACM, June, 60, 6, 84–90. Available from: https://cacm.acm.org/magazines/2017/6/217745-imagenet-classification-with-deep-convolutional-neural-networks/fulltext (archived at https://perma.cc/763N-LHFY) [Last accessed 10 July 2020]

Krishevsky, A., Sutskever, I., and Hinton, G.E. (2012) ImageNet classification with deep convolutional neural networks. Available from: https://papers.nips.cc/paper/4824-imagenet-classification-with-deep-convolutional-neural-networks.pdf (archived at https://perma.cc/6DNJ-J6AH) [Last accessed 10 July 2020]

Kritzinger, W., Steinwender, A., Lumetzberger, S., and Sihn, W. (2018) Impacts of additive manufacturing in value creation system, *Procedia CIRP*, 72, 1518–1523. Available from: https://doi.org/10.1016/j.procir.2018.03.205 (archived at https://perma.cc/T8D7-9CVR) [Last accessed 5 May 2020]

Kuhn, T. (1962) *The Structure of Scientific Revolutions*, Chicago, The University of Chicago Press.

Kutler, A. and Serbee, D. (2019) Top policy trends 2020: the data revolution has come to the policy arena. PwC.com. Available from: www.pwc.com/us/en/library/risk-regulatory/strategic-policy/top-policy-trends.html (archived at https://perma.cc/3B9Q-7KER) [Last accessed 10 June 2020]

Lorek, L. (2019) Paradromics moved from Silicon Valley to Austin and is creating a brain modem, 4 February. Available from: http://siliconhillsnews.com/2019/02/04/paradromics-moved-from-silicon-valley-to-austin-and-is-creating-a-brain-modem/ (archived at https://perma.cc/755N-QJ6D) [Last accessed 7 August 2020]

Law (2020) Restrictions on genetically modified organisms: European Union, Library of Congress Law, Available from: www.loc.gov/law/help/restrictions-on-gmos/eu.php (archived at https://perma.cc/F95X-5K6Z) [Last accessed 8 July 2020]

Leopold, G. (2019) Emerging AI business model promotes distributed ML, Enterprise.ai, 6 December. Available from: www.enterpriseai.news/2019/12/06/emerging-ai-business-model-promotes-distributed-ml/ (archived at https://perma.cc/ZAK5-C26W) [Last accessed 15 June 2020]

Levi, P.J. et al. (2019) Macro-energy systems: toward a new discipline. Joule. Available from: DOI: 10.1016/j.joule.2019.07.017 [Last accessed 4 May 2020]

Lewis-Kraus, G. (2018) Inside the crypto world's biggest scandal, *Wired*, 19 June. Available from: www.wired.com/story/tezos-blockchain-love-story-horror-story/ (archived at https://perma.cc/8SKN-5P7U) [Last accessed 10 July 2020]

Li, X. (2003) Science and technology is not simply equal to sci-tech. *Genomics Proteomics Bioinformatics*, 1, 2, 87–89. Available from: doi:10.1016/s1672-0229(03)01012-x [Last accessed 4 May 2020]

Londre, L.S. (2007) Introducing the 9Ps of marketing, NinePs.com. Available from: www.nineps.com/ (archived at https://perma.cc/XU2V-9HN7) [Last accessed 15 June 2020]

Lynch, M. (2019) Boehringer Ingelheim and IBM bring blockchain to clinical trials, 14 February. Available from: www.outsourcing-pharma.com/article/2019/02/14/boehringer-ingelheim-and-ibm-bring-blockchain-to-clinical-trials (archived at https://perma.cc/NEB4-6RQW) [Last accessed 7 July 2020]

Lynch, S. (2017) Andrew Ng: why AI is the new electricity, Stanford GSB Insight, 11 March. Available from: www.gsb.stanford.edu/insights/andrew-ng-why-ai-new-electricity (archived at https://perma.cc/H8Y6-HQPL) [Last accessed 4 May 2020]

Lyons, K. (2020) US government officials using mobile ad location data to study coronavirus spread, The Verge, 29 March. Available from: www.theverge.com/2020/3/29/21198158/us-government-mobile-ad-location-data-coronavirus (archived at https://perma.cc/2MXM-V5J5) [Last accessed 11 June 2020]

Mac Carron, P., Kaski, K., and Dunbar, R. (2016) Calling Dunbar's numbers, *Social Networks*, 47, October, 151–155. Available from: www.sciencedirect.com/science/article/pii/S0378873316301095 (archived at https://perma.cc/KV8W-QAF7) [Last accessed 14 September 2020]

Maitland, A. and Steele, R. (2020) *INdivisible: Radically Rethinking Inclusion for Sustainable Business Results*, Young & Joseph Press.

Makkonen, T. (2013) Government science and technology budgets in times of crisis, *Research Policy*, 42, 3, April, 817–822. Available from: https://doi.org/10.1016/j.respol.2012.10.002 (archived at https://perma.cc/M9MV-4UQX) [Last accessed 9 June 2020]

Mandel, M. (2019) Why 2019 will be the year of the manufacturing platform, *Forbes*, 2 January. Available from: www.forbes.com/sites/michaelmandel1/2019/01/02/2019-the-year-of-the-manufacturing-platform/#7b42eed63688 (archived at https://perma.cc/RT2L-B68B) [Last accessed 4 May 2020]

MarketandMarkets (2020) Synthetic biology market – forecast to 2025. Available from: www.marketsandmarkets.com/Market-Reports/synthetic-biology-market-889.html (archived at https://perma.cc/JR6J-JRBL) [Last accessed 20 August 2020]

Markman, J. (2019) Government efforts to regulate 'big tech' will likely backfire; here's why, *Forbes*, 26 January. Available from: www.forbes.com/sites/jonmarkman/2019/01/26/government-efforts-to-regulate-big-tech-will-likely-backfire-heres-why/#291497b0318c (archived at https://perma.cc/TEY9-AVVX) [Last accessed 4 May 2020]

Marr, B. (2020) These 25 technology trends will define the next decade, *Forbes*, 20 April. Available from: www.forbes.com/sites/bernardmarr/2020/04/20/these-25-technology-trends-will-define-the-next-decade/#224059e229e3 (archived at https://perma.cc/TFJ9-D473) [Last accessed 10 June 2020]

Marx, K. (1990) *Capital: A Critique of Political Economy, Volume 1*, Penguin, New York, [1867].

May, K.T. (2013) Julie Taymor and other creative minds share how they work, TED Blog, 31 July. Available from: https://blog.ted.com/julie-taymor-and-other-creative-minds-share-how-they-start-their-incredibly-unique-works/ (archived at https://perma.cc/E52Q-4UEN) [Last accessed 25 July 2020]

McAfee, A. and Brynjolfsson, E. (2017) *Machine Platform Crowd*, Norton, New York.

McKinsey (2019) Refueling the innovation engine in vaccines, McKinsey.com, 8 May. Available from: www.mckinsey.com/industries/pharmaceuticals-and-medical-products/our-insights/refueling-the-innovation-engine-in-vaccines (archived at https://perma.cc/9ZBF-M7MR) [Last accessed 21 May 2020]

Mercer (2019) Blockchain in HR: interesting use cases for human resources, Voice on Growth, Mercer. Available from: https://voice-on-growth.mercer.com/en/articles/innovation/blockchain-for-human-resources.html (archived at https://perma.cc/97ML-C2YA) [Last accessed 6 July 2020]

Miglierini, G. (2018) 3D printing, a disruptive technology still lacking regulatory guidance in the EU, *Pharma World Magazine*, 19 July. Available from: www.pharmaworldmagazine.com/3d-printing-a-disruptive-technology-still-lacking-regulatory-guidance-in-the-eu/ (archived at https://perma.cc/SPD6-FV37) [Last accessed 12 July 2020]

Miller, R. (2019) Enterprise SaaS revenue hit $100B run rate, led by Microsoft and Salesforce, TechCrunch, 28 June. Available from: https://techcrunch.com/2019/06/28/synergy-research-finds-enterprise-saas-revenue-hits-100b-run-rate-led-by-microsoft-salesforce/ (archived at https://perma.cc/3QVM-5DYZ) [Last accessed 28 June 2020]

MIT Energy Initiative: Available from: https://energy.mit.edu/ (archived at https://perma.cc/AW9H-R3UK) [Last accessed 4 May 2020]

Mitra, R. (2019) Understand blockchain business models: complete guide, Blockgeeks. Available from: https://blockgeeks.com/guides/understand-blockchain-business-models/ (archived at https://perma.cc/TT3J-8Y77) [Last accessed 28 May 2020]

Mohammadi, A.K. (2019) How To Secure Your Network: Five Modern Alternatives to VPN, Bleepingcomputer.com, 17 June. Available from: www. bleepingcomputer.com/news/security/how-to-secure-your-network-five-modern-alternatives-to-vpn/ (archived at https://perma.cc/L3D3-HZ4U) [Last accessed 8 June 2020]

Mohsin, M. (2020) 10 TikTok statistics that you need to know in 2020 [Infographic], Oberlo.com, 3 July. Available from: www.oberlo.com/blog/tiktok-statistics (archived at https://perma.cc/MKH2-LZBJ) [Last accessed 13 September 2020]

Mollick, E. (2019) What the lean startup method gets right and wrong, *Harvard Business Review*, 21 October. Available from: https://hbr.org/2019/10/what-the-lean-startup-method-gets-right-and-wrong (archived at https://perma.cc/Z5JC-Y2U4) [Last accessed 15 June 2020]

Molteni, M. (2018) Ginkgo Bioworks is turning human cells into on-demand factories, *Wired*, 24 October. Available from: www.wired.com/story/ginkgo-bioworks-is-turning-human-cells-into-on-demand-factories/ (archived at https://perma.cc/GJ4H-S4NL) [Last accessed 9 June 2020]

Morton, H. (2019) Blockchain 2019 legislation. Available from: www.ncsl.org/research/financial-services-and-commerce/blockchain-2019-legislation.aspx (archived at https://perma.cc/EK3L-LNRR) [Last accessed 4 July 2020]

Mudry, A. and Mills, M. (2013) The early history of the cochlear implant: a retrospective, *JAMA Otolaryngol Head Neck Surg.*, 139, 5, 446–453. Available from: doi:10.1001/jamaoto.2013.293 [Last accessed 10 August 2020]

Murgia, M. (2019) Europe 'a global trendsetter on tech regulation', *FT*, 30 October. Available from: www.ft.com/content/e7b22230-fa32-11e9-a354-36acbbb0d9b6 (archived at https://perma.cc/6B29-32GK) [Last accessed 9 June 2020]

Musk, E. (2019) An integrated brain-machine interface platform with thousands of channels, BioRxiv, 17 July. Available from: doi: https://doi.org/10.1101/703801 (archived at https://perma.cc/VCZ2-FX42) [Last accessed 7 August 2020]

NASA (2012) Technology Readiness Level. 28 October 2012. Available from: www.nasa.gov/directorates/heo/scan/engineering/technology/txt_accordion1. html (archived at https://perma.cc/6RN2-9MQG) [Last accessed 8 April 2020]

NASA (2020) TRL Definitions. Available from: www.nasa.gov/pdf/458490main_TRL_Definitions.pdf (archived at https://perma.cc/5Z6K-JD63) [Last accessed 8 April 2020]

National Academies (2017) National Academies of Sciences, Engineering, and Medicine; National Academy of Medicine; National Academy of Sciences; Committee on Human Gene Editing: Scientific, Medical, and Ethical Considerations. Human Genome Editing: Science, Ethics, and Governance. Washington (DC): National Academies Press (US); 2017 Feb 14. Summary. Available from: www.ncbi.nlm.nih.gov/books/NBK447260/ (archived at https://perma.cc/8YFE-446H) [Last accessed 9 June 2020]

National Bureau of Economic Research (2011) Determinants of vaccine supply, Nber.org. Available from: www.nber.org/aginghealth/2011no3/w17205.html (archived at https://perma.cc/6SE7-ZFK2) [Last accessed 21 May 2020]

National Research Council (US) Committee on a Framework for Developing a New Taxonomy of Disease (2011) Toward Precision Medicine: Building a Knowledge Network for Biomedical Research and a New Taxonomy of Disease. Washington (DC): National Academies Press (US). Available from: www.ncbi.nlm.nih.gov/books/NBK92144/ (archived at https://perma.cc/8BTJ-KVSA) [Last accessed 7 April 2020]

Naziri, J. (2011) 15 influential innovations of the past 50 years, CNBC, 19 September. Available from: www.cnbc.com/2011/09/19/15-Influential-Innovations-of-the-Past-50-Years.html (archived at https://perma.cc/R3UK-KUF4) [Last accessed 2 June 2020]

NEET (2020) New Engineering Education Transformation (NEET). Available from: https://neet.mit.edu/ (archived at https://perma.cc/Y7S9-FA5E) [Last accessed 13 August 2020]

Newton, C. (2020) A sneaky attempt to end encryption is worming its way through Congress, The Verge, 12 March. Available from: www.theverge.com/interface/2020/3/12/21174815/earn-it-act-encryption-killer-lindsay-graham-match-group (archived at https://perma.cc/56G8-VL83) [Last accessed 11 June 2020]

NIH (2020) Francis Crick: biographical overview. Available from: https://profiles.nlm.nih.gov/spotlight/sc/feature/biographical-overview (archived at https://perma.cc/NH5N-SKBP) [Last accessed 25 July 2020]

Ning, S. and Wu, H. (2020) AI, machine learning and big data: China. Global Legal Insights. Available from: www.globallegalinsights.com/practice-areas/ai-machine-learning-and-big-data-laws-and-regulations/china (archived at https://perma.cc/8TH8-D6F9) [Last accessed 21 May 2020]

Nisbet, M. (2015) Inside America's science lobby: What motivates AAAS members to engage the public? The Conversation, 6 March. Available from: https://theconversation.com/inside-americas-science-lobby-what-motivates-aaas-members-to-engage-the-public-38065 (archived at https://perma.cc/66YW-WMCN) [Last accessed 9 June 2020]

OECD (1972) Interdisciplinarity: problems of teaching and research in universities, Paris, OECD. Available from: https://files.eric.ed.gov/fulltext/ED061895.pdf (archived at https://perma.cc/P4EA-NC6A) [Last accessed 21 July 2020]

O'Leary, R. (2019) A new era in 3-D printing, 16 May, MIT News. Available from: http://news.mit.edu/2019/new-era-3d-printing-0516 (archived at https://perma.cc/46YP-Z49D) [Last accessed 7 July 2020]

Orrick (2020) How to move to remote work and comply with U.S. privacy and cybersecurity laws, Orrick Trust Control, JD Supra.com, 25 March. Available from: www.jdsupra.com/legalnews/how-to-move-to-remote-work-and-comply-68839/ (archived at https://perma.cc/T6LZ-VCVP) [Last accessed 8 June 2020]

Osterwalder, A. and Pigneur, Y. (2010) Business Model Generation: A Handbook for Visionaries, Game Changers, and Challengers. Wiley, London.

Paglieri, J. (2005) The 3 places where Facebook censors you the most, CNN Money, 6 February. Available from: https://money.cnn.com/2015/02/06/technology/facebook-censorship/ (archived at https://perma.cc/46TS-DKDW) [Last accessed 11 June 2020]

Pappas, P. (2015) Top 10 cloud-based learning management systems for corporate training, Elearning Industry.com, 3 November [updated 2020]. Available from: https://elearningindustry.com/top-10-cloud-based-learning-management-systems-for-corporate-training (archived at https://perma.cc/2HRE-DLTM) [Last accessed 21 July 2020]

Parker, G.G., Alstyne, M.V., and Choudary, S.P. (2016) *Platform Revolution: How Networked Markets Are Transforming the Economy – And How to Make Them Work for You*, Norton, New York.

Parrish, S. (2020) Mental Models: the best way to make intelligent decisions (109 models explained), fs.blog. Available from: https://fs.blog/mental-models/#building_a_latticework_of_mental_models (archived at https://perma.cc/R7TN-DNQY) [Last accessed 13 August 2020]

Pauwels, E. (2013) Public understanding of synthetic biology, *BioScience*, 63, 2, February, 79–89. Available from: https://doi.org/10.1525/bio.2013.63.2.4 (archived at https://perma.cc/JR6G-598F) [Last accessed 20 August 2020]

Pentland, S. (2015) *Social Dynamics*, Penguin, New York.

Pinsker, J. (2020) Oh no, they've come up with another generation label, *The Atlantic*, 21 February. Available from: www.theatlantic.com/family/archive/2020/02/generation-after-gen-z-named-alpha/606862/ (archived at https://perma.cc/2EPM-W4HK) [Last accessed 13 August 2020]

Plastics (2018) Size and Impact Report, Plastics. Available from: www.plasticsindustry.org/sites/default/files/SizeAndImpactReport_Summary.pdf (archived at https://perma.cc/5YJG-DX4W) [Last accessed 8 July 2020]

Prabhakar, A. (2019) The merging of humans and machines is happening now. Available from: www.wired.co.uk/article/darpa-arati-prabhakar-humans-machines (archived at https://perma.cc/Y9WS-B2VL) [Last accessed 5 August 2020]

Priester, V. (2020) Polymath Mae Jemison encourages bolder exploration, collaboration, 25 February. Available from: https://researchblog.duke.edu/2020/02/25/polymath-mae-jemison-encourages-bolder-exploration-collaboration/ (archived at https://perma.cc/ZC85-W4EC) [Last accessed 25 July 2020]

Rayna, T. and Striukova, L. (2016) From rapid prototyping to home fabrication: how 3D printing is changing business model innovation, *Technological Forecasting and Social Change*, 102, January, 214–224. Available from: https://doi.org/10.1016/j.techfore.2015.07.023 (archived at https://perma.cc/JFL4-F6D9) [Last accessed 16 August 2020]

Reader, R. (2020) How open-source medicine could prepare us for the next pandemic, *Fast Company*, 30 April. Available from: www.fastcompany.com/90498448/how-open-source-medicine-could-prepare-us-for-the-next-pandemic (archived at https://perma.cc/6BL6-F3EA) [Last accessed 4 May 2020]

Reuters (2020) U.S. government limits exports of artificial intelligence software, Reuters.com, 3 January. Available from: www.reuters.com/article/us-usa-artificial-intelligence/u-s-government-limits-exports-of-artificial-intelligence-software-idUSKBN1Z21PT (archived at https://perma.cc/3NK5-6FAU) [Last accessed 22 May 2020]

Ries, E. (2011) *The Lean Startup*, Crown Business, New York.

Roberts, H., Cowls, J., Morley, J., Taddeo, M., Wang, V., and Floridi, L. (2019) The Chinese approach to artificial intelligence: an analysis of policy and regulation, SSRN, 1 September 1. Available from: http://dx.doi.org/10.2139/ssrn.3469784 (archived at https://perma.cc/GBK5-6U47) [Last accessed 22 May 2020]

RoboLaw (2014) Regulating emerging robotic technologies in Europe: robotics facing law and ethics. FP7.

Rochet, J.C. and Tirole, J. (2005) Two-sided markets: a progress report, November 29, 2005. Available from: http://idei.fr/sites/default/files/medias/doc/wp/2005/2sided_markets.pdf (archived at https://perma.cc/E58U-7DNQ) [Last accessed 15 June 2020]

Rogers, E.M. (1962) *Diffusion of Innovations*, Free Press of Glencoe, New York.

Rong, K., Patton, D., and Chen, W. (2018) Business models dynamics and business ecosystems in the emerging 3D printing industry, *Technological Forecasting and Social Change*, 134, September, 234–245. Available from: https://doi.org/10.1016/j.techfore.2018.06.015 (archived at https://perma.cc/B5CZ-FJ75) [Last accessed 16 August 2020]

Rosen, R. (2016) Why do Americans work so much? *The Atlantic*, 7 January. Available from: www.theatlantic.com/business/archive/2016/01/inequality-work-hours/422775/ (archived at https://perma.cc/4QZW-MV6P) [Last accessed 12 September 2020]

Salatino, A.A. (2020) The computer science ontology: a comprehensive automatically-generated taxonomy of research areas. Data Intelligence 2. Available from: www.mitpressjournals.org/doi/pdf/10.1162/dint_a_00055 (archived at https://perma.cc/8BCB-N7ML) [Last accessed 12 May 2020]

Sassen, S. (1991) *The Global City*, Princeton University Press, Princeton NJ.

Schaaf, T. (2020) In MedTech History, MedTech Strategist. 20 May. Available from: www.medtechstrategist.com/in-medtech-history (archived at https://perma.cc/MMM5-QTJA) [Last accessed 20 May 2020]

Schiavi, G.S. and Behr, A. (2018) Emerging technologies and new business models: a review on disruptive business models, *Innovation & Management Review*, 15, 4, 338–355. Available from: https://doi.org/10.1108/INMR-03-2018-0013 (archived at https://perma.cc/QL2V-DY5W) [Last accessed 20 May 2020]

Schön, D. (1983) *The Reflective Practitioner: How Professionals Think in Action*, Basic Books, New York.

Schumpeter, J.A. (1942) *Capitalism, Socialism, and Democracy*, Harper & Brothers, London.

Schweighart, V. (2020) What will the medtech sector look like under EU MDR law? MedCityNews, 23 January. Available from: https://medcitynews.com/2020/01/what-will-the-medtech-sector-look-like-under-eu-mdr-law/ (archived at https://perma.cc/3SMY-U26Q) [Last accessed 12 July 2020]

Understanding Science (2020) Science and technology on fast forward. UC Berkeley. Available from: https://undsci.berkeley.edu/article/0_0_0/whathassciencedone_03 (archived at https://perma.cc/Q5P8-ZX4D) [Last accessed 8 April 2020]

Science Daily (2019) New discipline proposed: macro-energy systems – the science of the energy transition, *Science Daily*, 19 August. Available from: www.sciencedaily.com/releases/2019/08/190819092959.htm (archived at https://perma.cc/D75J-R4EY) [Last accessed 5 May 2020]

Seck, H.H. (2017) Super SEALs: elite units pursue brain-stimulating technologies, Military.com, 2 April. Available from: www.military.com/daily-news/2017/04/02/super-seals-elite-units-pursue-brain-stimulating-technologies.html (archived at https://perma.cc/T9CR-PY7L) [Last accessed 7 August 2020]

Segran, E. (2019) Yuval Noah Harari: humans are on the verge of merging with machines, *Fast Company*, 19 September. Available from: www.fastcompany.com/90373620/yuval-noah-harari-humans-are-on-the-verge-of-merging-with-machines (archived at https://perma.cc/3V62-PKDA) [Last accessed 7 August 2020]

Shaer, M. (2014) Is this the future of robotic legs? *Smithsonian Magazine*. Available from: www.smithsonianmag.com/innovation/future-robotic-legs-180953040/?page=2 (archived at https://perma.cc/TGT9-E384) [Last accessed 10 August 2020]

Sherman, J. (2020) Oh sure, big tech wants regulation—on its own terms, *Wired*, 20 January. Available from: www.wired.com/story/opinion-oh-sure-big-tech-wants-regulationon-its-own-terms/ (archived at https://perma.cc/U4VB-K9YH) [Last accessed 9 June 2020]

Sinfield, J.V., Calder, E., McConnell, B., and Colson, S. (2011) How to identify new business models, *MIT Sloan Management Review*, 21 December. Available from: https://sloanreview.mit.edu/article/how-to-identify-new-business-models/ (archived at https://perma.cc/D7JN-SWEN) [Last accessed 15 June 2020]

Singer, P. (2004) Federally supported innovations: 22 examples of major technology advances that stem from federal research support, ITIF.org, The Information Technology and Innovation Foundation, February. Available from: www2.itif.org/2014-federally-supported-innovations.pdf (archived at https://perma.cc/T998-2KU6) [Last accessed 11 June 2020]

Singh, J. (2019) Merging with AI: how to make a brain-computer interface to communicate with Google using Keras and OpenBCI, Medium.com, 5 September. Available from: https://towardsdatascience.com/merging-with-ai-how-to-make-a-brain-computer-interface-to-communicate-with-google-using-keras-and-f9414c540a92 (archived at https://perma.cc/J8EC-K5BL) [Last accessed 7 August, 2020]

Smart Dubai (2020) Dubai Blockchain strategy. Available from: www.smartdubai.ae/initiatives/blockchain# (archived at https://perma.cc/3589-LWLN) [Last accessed 10 July 2020]

Smit, S., Tacke, T., Lund, S., Manyika, J., and Thiel, L. (2020) The future of work in Europe, McKinsey Global Institute, Discussion paper, 10 June. Available from: www.mckinsey.com/featured-insights/future-of-work/the-future-of-work-in-europe# (archived at https://perma.cc/AV2B-LNT2) [Last accessed 6 July 2020]

Stanford (2005) 'You've got to find what you love,' Jobs says, *Stanford News*, 14 June. Available from: https://news.stanford.edu/2005/06/14/jobs-061505/ (archived at https://perma.cc/86EM-L7UK) [Last accessed 25 July 2020]

Steinbuch, Y. (2018) Jeff Bezos tells employees that Amazon 'is not too big to fail, New York Post, 16 November. Available from: https://nypost.com/2018/11/16/jeff-bezos-tells-employees-that-amazon-is-not-too-big-to-fail/ (archived at https://perma.cc/97BU-4W7L) [Last accessed 14 August 2020]

Sterman, J.D. (1987) The economic long wave: theory and evidence. In: Vasko T. (ed) *The Long-Wave Debate*, Springer, Berlin, Heidelberg. Available from: https://doi.org/10.1007/978-3-662-10351-7_11 (archived at https://perma.cc/5SZ4-228Z) [Last accessed 15 August 2020]

Sternberg, R. (2020) Balance theory of wisdom. Available from: www.robertjsternberg.com/wisdom (archived at https://perma.cc/2WR8-8FHH) [Last accessed 18 August 2020]

Stillman, J. (2017) Here's Elon Musk's secret for learning anything faster, Inc., 23 August. Available from: www.inc.com/jessica-stillman/heres-elon-musks-secret-for-learning-anything-fast.html (archived at https://perma.cc/838D-YHFP) [Last accessed 10 July 2020]

Stoddard, J., Drucker, C., and Brown, N. (2007) 21st century vaccines – a development renaissance, Drug Discovery World (DDW Online). Available from: www.ddw-online.com/enabling-technologies/p92830-21st-century-vaccines-a-development-renaissance.html (archived at https://perma.cc/Y9KS-SY72) [Last accessed 21 May 2020]

Subrahmanian, E. and Reich, Y. (2020) *Why We Are Not Users: Dialogues, Diversity and Design*, MIT Press, Cambridge, MA.

Sweat, S. (2016) Segway accidents in California, *National Law Review*, VI, 340, 5 December. Available from: www.natlawreview.com/article/segway-accidents-california (archived at https://perma.cc/D7GU-5ZZT) [Last accessed 26 June 2020]

Synthego (2019) The power of synthetic biology: 25 thought leaders opine, Synthego, 17 January. Available from: www.synthego.com/blog/synthetic-biology-applications (archived at https://perma.cc/4EUS-K336) [Last accessed 12 July 2020]

Tank, A. (2020) Why the world needs deep generalists, not specialists, JotForm, 16 July. Available from: www.jotform.com/blog/the-world-needs-polymaths/ (archived at https://perma.cc/ND2L-AXS8) [Last accessed 21 July 2020]

Tank, A. (2019) The era of the specialist is over, *Entrepreneur*, 21 February. Available from: www.entrepreneur.com/article/327712 (archived at https://perma.cc/D59V-D3FP) [Last accessed 25 July 2020]

Tesla, N. (1919) My inventions, *Electrical Experimenter magazine*. Available from: www.tfcbooks.com/special/my_inventions_index.htm (archived at https://perma.cc/9YLS-3PUX) [Last accessed 25 July 2020]

Thalassemia.org (2019) First Beta Thalassemia patient treated with CRISPR/Vertex CTX001 Now Transfusion-Independent, 20 November. Available from: www.thalassemia.org/first-beta-thalassemia-patient-treated-with-crispr-vertex-ctx001-now-transfusion-independent/ (archived at https://perma.cc/PHS7-QY2Q) [Last accessed 5 May 2020]

The Manufacturing Institute (2018) Manufacturing industry faces unprecedented employment shortfall, 14 November. Available from: www.themanufacturinginstitute.org/press-releases/manufacturing-industry-faces-unprecedented-employment-shortfall-2-4-million-skilled-jobs-projected-to-go-unfilled-according-to-deloitte-and-the-manufacturing-institute/ (archived at https://perma.cc/24KQ-XN8R) [Last accessed 12 September 2020]

Tolman, E.C. (1948) Cognitive maps in rats and men, *Psychol. Rev*, 55, 189–208. Available from: DOI: 10.1037/h0061626 [Last accessed 7 August 2020]

Trustradius (2020) Corporate learning management systems overview. Available from: www.trustradius.com/corporate-learning-management (archived at https://perma.cc/Y9NM-LAWG) [Last accessed 21 July 2020]

Tsui, K. (2020) Transhumanism: meet the cyborgs and biohackers redefining beauty, CNN, 27 May. Available from: www.cnn.com/style/article/david-vintiner-transhumanism/index.html (archived at https://perma.cc/GAG5-5GGC) [Last accessed 7 August 2020]

Uenlue, M. (2018) Amazon business model: the ultimate overview, Innovation Tactics, 14 December. Available from: https://innovationtactics.com/amazon-business-model-ultimate-overview/#Fundamental-business-model-principles (archived at https://perma.cc/9KBB-5SZM) [Last accessed 28 May 2020]

UNDESA (2018) 2018 UN E-Government Survey, 19 July. Available from: www.un.org/development/desa/publications/2018-un-e-government-survey.html (archived at https://perma.cc/MS5A-Y62D) [Last accessed 9 June 2020]

Undheim, K. and Aasebø, A. (2019) *Passionistas – Women of Influence*, Grapes, Oslo.

Urban, T. (2017) Neuralink and the brain's magical future, wait but why, 20 April. Available from: https://waitbutwhy.com/2017/04/neuralink.html (archived at https://perma.cc/6555-HJVR) [Last accessed 7 August 2020]

U.S. Census Bureau (2017) Average one-way commuting time by metropolitan areas, U.S. Census Bureau, 7 December. Available from: www.census.gov/library/visualizations/interactive/travel-time.html (archived at https://perma.cc/KT27-B868) [Last accessed 21 May 2020]

Valamis (2020) The definitive guide to microlearning. Available from: www.valamis.com/blog/the-definitive-guide-to-microlearning (archived at https://perma.cc/ZDF4-E8KQ) [Last accessed 21 July 2020]

Vance, A. (2017) *Elon Musk: Tesla, SpaceX, and the Quest for a Fantastic Future*, Ecco.

Waldert, S. (2016) Invasive vs. Non-Invasive Neuronal Signals for Brain-Machine Interfaces: Will One Prevail? Frontiers in Neuroscience, Vol. 10, p. 295. Available from: www.frontiersin.org/article/10.3389/fnins.2016.00295 (archived at https://perma.cc/UZK2-KDML) [Last accessed 7 August 2020]

Warwick, K. (2020) Available from: Project Cyborg 1.0 www.kevinwarwick.com/project-cyborg-1-0/ (archived at https://perma.cc/YK8T-QVY2) [Last accessed 7 August 2020]

Wasmer, M. (2019) Roads forward for European GMO policy—uncertainties in wake of ECJ judgment have to be mitigated by regulatory reform, Frontiers of Bioengineering and Biotechnology, 05 June 2019. Available from: https://doi.org/10.3389/fbioe.2019.00132 (archived at https://perma.cc/G9M4-9P2K) [Last accessed 7 July 2020]

Weber, M. (1922) *Economy and Society*, University of California Press, Berkeley, CA.

Weber, M. (2005) Remarks on technology and culture, *Theory, Culture & Society*, 22, 4, August. Available from: https://doi.org/10.1177/0263276405054989 (archived at https://perma.cc/CJL3-B7NU) [Last accessed 2 June 2020]

Wee, S.L. and Mozur, P. (2019) China's genetic research on ethnic minorities sets off science backlash, *New York Times*, 4 December. Available from: www.nytimes.com/2019/12/04/business/china-dna-science-surveillance.html (archived at https://perma.cc/SWW4-BWJ9) [Last accessed 8 July 2020]

Weidner, J.B. (2020) How & why Google Glass failed, Investopedia, 8 March. Available from: www.investopedia.com/articles/investing/052115/how-why-google-glass-failed.asp (archived at https://perma.cc/4M69-3MGV) [Last accessed 28 June 2020]

Weisskircher, M. (2019) New technologies as a neglected social movement outcome: the case of activism against animal experimentation, *Sociological Perspectives*, 62, 1, 59–76. Available from: https://doi.org/10.1177/0731121418788339 (archived at https://perma.cc/25M2-7ADR) [Last accessed 2 June 2020]

Weller, M. (2018) Twenty years of edtech, *EduCause Review*, 2 July. Available from: https://er.educause.edu/articles/2018/7/twenty-years-of-edtech (archived at https://perma.cc/EDG8-C2U8) [Last accessed 5 May 2020]

West, D.W. and Allen, J.R. (2020) *Turning Point: Policymaking in the Era of Artificial Intelligence*, Brookings Institution Press, Washington, DC.

West, D.W. and Allen, J.R. (2018) How artificial intelligence is transforming the world, Brookings Institution, 24 April. Available from: www.brookings.edu/research/how-artificial-intelligence-is-transforming-the-world/ (archived at https://perma.cc/3XCP-4DZ9) [Last accessed 10 July 2020]

Wiles, J. (2019) Early adopters are already using blockchain-inspired approaches in certain HR areas. These use cases offer HR leaders a glimpse of what's to come, 27 August. Available from: www.gartner.com/smarterwithgartner/5-ways-blockchain-will-affect-hr/ (archived at https://perma.cc/Q9WH-2U6J) [Last accessed 6 July 2020]

Wilson, M. (2020) Exclusive: Segway, the most hyped invention since the Macintosh, ends production, *Wired*, 23 June. Available from: www.fastcompany.com/90517971/exclusive-segway-the-most-hyped-invention-since-the-macintosh-to-end-production (archived at https://perma.cc/VBK6-EFA3) [Last accessed 26 June 2020]

Wittenberg-Cox, A. (2020) 5 Economists redefining... everything. Oh yes, and they're women, *Forbes*, 31 May. Available from: www.forbes.com/sites/avivahwittenbergcox/2020/05/31/5-economists-redefining-everything–oh-yes-and-theyre-women/#120e918d714a (archived at https://perma.cc/M9CV-MGM6) [Last accessed 2 June 2020]

Wladawsky-Berger, I. (2015) The rise of the T-shaped organization, *The Wall Street Journal*, 18 December. Available from: https://blogs.wsj.com/cio/2015/12/18/the-rise-of-the-t-shaped-organization/ (archived at https://perma.cc/AS2F-83YC) [Last accessed 25 July 2020]

Wohlers, T. and Gornet, T. (2014) History of additive manufacturing, Wohlers Report. Available from: www.wohlersassociates.com/history2014.pdf (archived at https://perma.cc/3QEY-UCXF) [Last accessed 7 July 2020]

Wopata, M. (2019) The leading industry 4.0 companies 2019, IoT Analytics, Available from: https://iot-analytics.com/the-leading-industry-4-0-companies-2019/ (archived at https://perma.cc/VY96-C3CZ) [Last accessed 4 May 2020]

Working With McKinsey (2013) T-shaped problem-solving at McKinsey and 3 reasons why it's preferred, 12 January. Available from: http://workingwithmckinsey.blogspot.com/2013/01/t-shaped-problem-solving-at-mckinsey.html (archived at https://perma.cc/WUJ6-GJB5) [Last accessed 20 July 2020]

World Bank (2007) Review of the Dutch Administrative Burden Reduction Programme, World Bank. Available from: www.doingbusiness.org/content/dam/doingBusiness/media/Special-Reports/DB-Dutch-Admin.pdf (archived at https://perma.cc/BZ84-TK4J) [Last accessed 10 June 2020]

Wu, J. (2020) Cybersecurity when it comes to remote work means zero trust, *Forbes*, 18 March. Available from: www.forbes.com/sites/cognitiveworld/2020/03/18/cybersecurity-when-it-comes-to-remote-work-means-zero-trust/#717aa59bc5b1 (archived at https://perma.cc/U776-BFYH) [Last accessed 8 June 2020]

XTRD (2020) Top 100 blockchain and crypto influencers on Twitter to follow, XTRD. Available from: https://xtrd.io/top-100-blockchain-and-crypto-influencers-on-twitter-to-follow/ (archived at https://perma.cc/3S53-VCGM) [Last accessed 10 July 2020]

Yates, J. and Murphy, C.N. (2019) *Engineering Rules: Global Standard Setting Since 1880*, Johns Hopkins University Press, Baltimore, MD.

Yoshi, R. (2020) How can blockchain be implemented in the life sciences ecosystem? 11 February. Available from: www.technologynetworks.com/informatics/articles/how-can-blockchain-be-implemented-in-the-life-sciences-ecosystem-330614 (archived at https://perma.cc/25KF-69SZ) [Last accessed 7 July 2020]

Zuora (2020) The subscription economy. Available from: www.zuora.com/vision/subscription-economy/ (archived at https://perma.cc/8ZEA-W9LA) [Last accessed 13 September 2020]

INDEX

Italics denote a table or figure.

CPSIA information can be obtained
at www.ICGtesting.com
Printed in the USA
LVHW021705250221
679549LV00007B/2